Modern Sawmill Techniques

Volume 6

Also published by Miller Freeman Publications for the forest products industry:

Lumber Manufacturing
The Design and Operation of Sawmills and Planer Mills
by Ed M. Williston

Plywood Manufacturing Practices
by Richard F. Baldwin

European Sawmill Techniques
Proceedings from the First International Sawmill Seminar, Munich, June 1974.
Available in English, French, and German

Nordic and North American Sawmill Techniques
Proceedings from the Second International Sawmill Seminar, Jönköping, Sweden, June 1975

Modern Plywood Techniques
Volumes in the Plywood Clinic Library

- **Volume 1**
  Proceedings of the first Plywood Clinic, Portland, Oregon, February 1974
- **Volume 2**
  Proceedings of the second Plywood Clinic, New Orleans, Louisiana, November 1974
- **Volume 3**
  Proceedings of the third Plywood Clinic, Portland, Oregon, March 1975
- **Volume 4**
  Proceedings of the fourth Plywood Clinic, Portland, Oregon, March 1976

Logging Practices
Principles of Timber Harvesting Systems
by Steve Conway

Timber Cutting Practices
by Steve Conway

Sawmill Techniques for Southeast Asia
Proceedings of the First Southeast Asia Sawmill Seminar, Singapore, June 1975

Proceedings of the World Consultation on Wood-Based Panels
Held in New Delhi, India, February 1975.
Available in English, French, and Spanish

Modern Sawmill Techniques
Other volumes in the Sawmill Clinic Library

- **Volume 1**
  Proceedings of the first Sawmill Clinic, Portland, Oregon, February 1973
- **Volume 2**
  Proceedings of the second Sawmill Clinic, New Orleans, Louisiana, November 1973
- **Volume 3**
  Proceedings of the third Sawmill Clinic, Portland, Oregon, February 1974
- **Volume 4**
  Proceedings of the fourth Sawmill Clinic, New Orleans, Louisiana, November 1974
- **Volume 5**
  Proceedings of the fifth Sawmill Clinic, Portland, Oregon, March 1975

Business Management for Sawmills & Plywood Mills
Proceedings of the Sawmill & Plywood Clinic Business Management Workshop
Portland, Oregon, March 4, 1976

Modern Sawmill Techniques

Volume 6: Proceedings of
the sixth Sawmill Clinic,
Portland, Oregon, March 1976

Edited by Vernon S. White

Volume 6 of the Sawmill Clinic Library
A FOREST INDUSTRIES book

Library of Congress Catalog Card Number 73-88045
International Standard Book Number 0-87930-052-3

Table of Contents

	Foreword by the editor	7
	Clinic authors & chapter summaries	10
1	Dry-land log handling: dump to infeed deck By Scott Corbett, Jr.	35
2	Upgrading log bucking and sorting methods By Erwin C. Bahlburg	48
3	Automation: keeping it practical and running By Amos A. Horner	58
4	Modifying a mill to fit today's logs By Tom Fleming	73
5	A way to keep your small-log side simple By Ray C. Swanson	86
6	Attention to detail can reduce lost-time dollars By David H. King	100
7	Control logic functions inexpensively and precisely By Robert P. Lewis	141
8	Raise sideboard recovery with computerized edger By Chris Vadnais	154
9	Points in selecting a wastewood chipper By Fred Veuger	162
10	Improving drive efficiency with synthetic flat belts By Greg Baird	173

Table of Contents

11 Building a practical quality control program 186
By Dean Huber, George H. Sisterhenm,
Robert L. Mardock, James L. Russell, Robert W. Nix
& Frank John Banducci, Jr.

12 Compressed air—without oil-coolant worries 229
By John G. Greenwood

13 Opening the flow from bulk storage containers 236
By Rick Bickers

14 Dry kiln design and woodwaste recovery 242
By Jack L. Bachrich

15 Optimum use of your top resource—people 258
By Jay Gruenfeld

16 Human factors in engineering design 265
By Joe G. Sublett

17 The impact of land-use planning 273
By V. M. "Whitey" Howard & William Lesh

18 The national forests: some neglected costs 288
By Barney Dowdle

Exhibitors, 1976 Sawmill & Plywood Clinic, Portland, Oregon 299

Index 301

Foreword by the editor

Ideas. Ideas to increase grade and volume recovery, to speed production. Ideas to reduce downtime and maintenance, to cut energy consumption and labor. These are the backbone of the Sawmill Clinics and hence the backbone of this book.

The exchange of valuable ideas among sawmill operators and their superintendents continues to evidence an open spirit of cooperation which brings the industry great credit. And the industry's ability to get the most from its raw material with the least use of energy and labor truly benefits the economies of the two nations which take part in the Clinics.

In my opinion, one of the most valuable Clinic presentations came this year. It was David H. King's bagful of ideas to solve sawmill problems. Head of a British Columbia sawmill consulting firm, King has accumulated a lifetime's experience with the nagging problems that cause hangups and downtime and poor performance by both worker and machine. Much of his knowledge was gained heading construction, engineering, and production for MacMillan Bloedel sawmills in the big-timber country of coastal British Columbia and Vancouver Island. King knows what it takes to keep things moving smoothly and he proved a prime example of mill men willing to share their ideas with the entire industry. His talk was chock-full of suggestions, instructions, and hints—more than many could remember. But here it is, complete with sketches and charts, in this book.

A major panel on the practical side of quality control was organized by Dean Huber, formerly a private consultant who now advises on sawmill efficiency for the U.S. Forest Service. Five mill men presented their experiences from the varying viewpoints of establishing management objectives, machinery modification and its impact, motivating the work force, and so forth. Quality control and grade/volume recovery really cannot be separated, so the potential impact of the panel's ideas on a mill's profit could be enormous.

But perhaps the highlight of the 1976 Clinic was Amos Horner's story of how he has kept automation simple and running at his

Tomco mills in Linn County, Oregon. Again and again, Horner emphasizes the importance of people. From conceiving the best system to convert available raw material to the most profitable products, through machinery selection with the help of the men who must run and maintain it, through the commitment which must be had from planning consultants and machinery installers, Horner explains his experience in getting his crew committed to making the mill GO! And almost as an afterthought, he tells the reader not to expect automation to reduce the size of the crew. More production per person . . . higher quality production per person . . . a safer, more interesting, more profitable job per person. But not fewer people.

Energy costs remain uppermost in the minds of many mill men. Jack Bachrich's discussion of how to do the most work with the Btus contained in mill waste certainly was a Clinic bell ringer if measured by the questions and discussion it generated. Since the days when whaling ships were driven by sail, no other industry has had equal opportunity to supply its energy needs from the residuals of its manufacturing process. The Austrian-born consultant who came to Canada via Britain has very practical suggestions for getting the energy where it is needed at the least cost.

The surging requirements for capital form a barrier that the sawmill operator must hurdle if he is to survive. Rising log costs have brought on the need for sophisticated electronic gear to improve recovery. But scanners and computers and high-recovery automatic machinery are also themselves vastly more expensive. For the operator trained as a mill man, finance may be an Alice-in-Wonderland world. To help him, special business topics were added to the 1976 Portland Sawmill Clinic. And two very hot current topics also were included: the Monongahela decision's potentially disastrous restriction on the flow of national forest timber and conspiracy to limit bidding on federal timber. Three distinguished Portland attorneys developed those themes. Because of the current, quick-changing nature of the business topics, they have been printed separately and distributed to those who attended the Clinic. The business sessions were conducted by Herb Lambert, editor of *Forest Industries* magazine.

The Clinic's Sawmill & Plywood Equipment Show has evolved into a major trade show—the principal U.S. show outside the

South. It began with the use of only 48 small booths at the First Clinic in 1973 and has grown to about 300 booth spaces. But more interesting in 1976 was the inclusion of really large, expensive items of sawmill equipment, such as high-strain, thin-kerf headrig bandmills, headrig carriages, planers, and sorting systems. And those attending the Clinic appear to spend an increasing portion of their time in the equipment arena.

Production of a book such as this requires the effective efforts of many. I would like to add my thanks to those of Miller Freeman Publications to the people whose participation in the Clinic made the content possible and to those who helped us shape their information into this volume.

Vernon S. White
Editor
Portland, Oregon
July 1976

Clinic Authors & Chapter Summaries

CHAPTER 1

DRY-LAND LOG HANDLING: DUMP TO INFEED DECK

By Scott Corbett, Jr.

SCOTT CORBETT, JR. is president of Howard-Cooper Equipment Co., one of the leading equipment suppliers to the wood products industry. Under his leadership, the Portland, Oregon, firm has expanded nationwide in heavy materials handling. Born in China where his father was stationed on business, he majored in business and political science at the University of Oregon and he has a Management Clinic A.M.A. He has been in machinery sales and marketing since World War II, working with logging and construction equipment and with intermodal and heavy lift fork trucks. He was with Nelson Equipment Co. from 1945 to 1970 and has been with Howard-Cooper since. Active in industry affairs, he is past president of the Oregon Machinery Dealers and the National Railroad Piggyback associations and was chairman of the Suppliers committee for the Pacific Logging Congress from 1971 through 1975.

CHAPTER SUMMARY: The evolution of dry-land log handling was hastened by rising costs of logs and labor. Rising cost of land was incidental.

The rise in log cost required that the best return be obtained from processing it. This, in turn, necessitated sorting. Improving markets for residues—especially chips—required sorting to species.

Labor costs made it inefficient to have a pond man sort to species, then diameter, then grade. More rapid production in the mill required an increase in mill deck volume. Meantime, smaller logs required handling more pieces. So the value of dry-land handling was enhanced.

Pettibones with the Lature forward extension (10,000- to 30,000-pound capacities) were the forerunners of dry-land handling with forklifts. R. G. LeTourneau produced the first large production-model forklift for dry-land log handling. The Raygo-Wagner Lumberjack was developed to release trucks faster, to keep the truckload as a unit until scaled, and to permit decks to be stacked higher. Crane and heelboom development kept pace.

The factors which brought about dry-land log handling are accelerating, making effective use of the yard even more vital.

The yard should be a separate profit center. Formerly, logger (or trucker) responsibility ended at the log dump, but the mill's did not begin until the infeed–leaving the yard a no-man's land. But it has great potential for savings and for increasing production.

Logs arrive either by truck or railcar. Unloading systems including winches, cranes, grapples, and lift trucks. Lift trucks range to 70,000-pound capacities and are dominated by LeTourneau and Raygo-Wagner and include Kenworth Dart. Smaller units will handle most truckloads. Surface conditions may make a four-wheel-drive unit necessary. Surface has a direct bearing on handling speed. Advances in scaling and sorting systems are described.

CHAPTER 2 UPGRADING LOG BUCKING AND SORTING METHODS

By Erwin C. Bahlburg

ERWIN C. "ERV" BAHLBURG is assistant general manager for wood products at Publishers Paper Co., Portland, Oregon. His responsibility includes all Publishers' sawmills and plywood plants. Prior to joining the firm two years ago, Bahlburg was with Cascadia Lumber Co., at Toledo, Oregon, for 22 years as a manager of sawmill operations there. Previously, he had spent 12 years with Keith Brown Building Supply Co., at Salem, Oregon. There, he was plant manager of the remanufacturing factory, which produces complete home building products and manufacturers prefabricated houses and prefab components. Bahlburg studied structural architecture at Oregon State University for two years, then transferred to the University of Oregon to finish in structural architecture and design.

CHAPTER SUMMARY: To handle increased production and barking for its Molalla, Oregon, sawmill, Publishers Paper Co. chose the Totem-Sateko log sorting system. The mill had a small pond and two decks. Production was less than 200,000 board feet per shift with band headrig and quad. With a computer, the Totem system, and a chain directly to the mill, production has increased to 260,000 to 320,000 per shift. Bucked log piece count averages 2,600, which also supplies the second shift. The other decks still can supply the mill if there is a breakdown.

The system's infeed deck is 12 to 14 feet above ground. A crane infeeds, and there is a chop saw following the deck. Bahlburg would prefer a lower ramp and chain with a basement containing refuse conveyors. It is hard for the crane operator to see how he is dropping logs onto the chain. The crane should have been placed at the end of the infeed chain. The 30-inch Soderhamn Cambio barker handles 4-inch to 27-inch logs.

The computerized bucking station measures length in 1-inch increments, measures diameter, and instructs the operator how the log should be bucked. It is computed for best market prices and information is kept current by changing computer cards. Average profit return has increased considerably. Operator overrides average only 30 to 45 per shift. Average length of bucked logs is greater. Daily printout summarizes bucking activity.

The Totem system then centers the log on four chairs and the operator signals whether it should go to the headrig (13 inches and up) or to the quad (smaller logs). He also indicates the bay. When the log arrives at the bay, the track collapses, jolting the log into rolling off as the chairs slant. The conveyor should be protected from being struck by loaders. The system will take any sweep or crook and is reliable in bay selection.

Activation of a solenoid lifts a trip lever above the rail. When the chair hits the lever, it collapses the rail, tilting the chairs. The first chair resets the rail.

CHAPTER 3 AUTOMATION: KEEPING IT PRACTICAL AND RUNNING

By Amos A. Horner

AMOS A. HORNER is the president and owner of Tomco, Inc., Sweet Home and Cascadia, Oregon, and the All-American Stud Co., at Springfield, Oregon. He has a long history of work with sawmills in which he mastered the diverse trades of logger, welder, carpenter, industrial designer and engineer, and business manager. The insight he has gained and his persistence are shown in the name he chose for his firm—Tomco—which stands for "Try Once More Co." Very active in the industry, he is president of North West

Timber Assn. and is a board member of West Coast Lumber Inspection Bureau, the Oregon Forest Industries Council, and the Oregon State Forest Research Advisory Committee. Born in Portland, Oregon, Horner had his schooling there and through correspondence but refers to himself as a graduate of "the school of hard knocks." He has traveled to Japan and elsewhere in the Orient to study log export problems and lumber export opportunities.

CHAPTER SUMMARY: Astronomical stumpage prices and timber scarcity have many mills almost to their economic knees. The Forest Service, plagued by environmentalists, is doing an increasingly poor job of managing our resource.

To meet the challenge, Tomco found automation *can* be practical. It *can* run.

But it did not work the first time, an experience shared by many mills. Some scanning devices, for instance, have been on line for two years and are still not dependable. But Tomco plunged ahead to try to get maximum yield with more precise machines, many computerized, with automatic control of charging, setting, feeding, sorting, and so forth.

In automating any function, involve your people early. Machine performance will depend upon your staff. Second, hire professional help. Get commitments from it. Research what is running in other mills. Visit mills. Get ideas on paper.

Mills no longer can afford mistakes in cutting. While the firm's stud mill at Cascadia, Oregon, cuts only logs under 12 inches, size variation requires that they be cut to different patterns to get maximum yield. A sawyer cannot make the decision accurately enough to maximize yield or quickly enough to maximize production. It is essential that the scanner work precisely—repeatedly.

Waste time must be minimized. Keep the wood in the saw! Keep systems simple.

If a technically trained person must be added, find and train him early. Failure to challenge your people and to get commitments from them will result in lost production and high maintenance. Automation will not reduce the number of people needed, but it will make the people who are used more productive.

A room for a computer should be on a solid foundation. Its technician should be able to see the machines on the mill floor to check the effect of his changes upon their performance.

Monitor the average number of seconds per log at each station. Have the operator account for all downtime exceeding 25 seconds. Analyze each operation.

CHAPTER 4

MODIFYING A MILL TO FIT TODAY'S LOGS

By Tom Fleming

TOM FLEMING is superintendent of lumber operations for American Forest Products Corp., a subsidiary of the Bendix Corp. He is engaged in planning changes at the Martell, California, sawmill to best fit the mill to a change in the log mix toward smaller sizes. Born in San Diego, California, Fleming obtained a bachelor's degree from the University of Washington in forest products. He joined Wheeler-Osgood Co., becoming assistant director of industrial relations, moved to Tabor City Lumber as sales and technical representative, and then to Glencoe Forest Products as assistant to the owners. He has a special interest in industrial photography. He has been secretary-treasurer of the Sawmill Club of Northern California for 23 years and is active in the Forest Products Research Society.

CHAPTER SUMMARY: The problem facing American Forest Products at Martell was how to change a large-log mill to fit a decreasing log size most advantageously. This is true throughout the sawmill industry, and many approaches have been taken toward solving it. Smaller logs mean lower recovery factors and more headrig time in relation to production.

A timberland study helped, with a breakdown of the diameter classes which would be coming to the mill.

Of pieces reaching the trimmer, 44.9 percent come from the resaw and 30.5 percent from the 6-inch edger, 21.2 percent from the big edger, 2.1 percent direct from the double-cut band, and 1.3 percent direct from the single-cut band headrig.

A single cutoff and barking system serves both the sawmill and plywood mill, but the conveyor to the sawmill requires a log to travel six minutes to the headrig. There are two 9-foot Filer & Stowell headrigs—a single-cut and a double-cut—with an 8-inch Portland Ironworks and a six-inch Klamath edger. Both headrigs have Beloit Passavant slabbers, 24- and 36-inch. There is an Irvington-Moore 20-foot trimmer and a Moore-Oregon 24-tray sorter.

Installation of a separate small-log side would be redundant in light of the decreasing log size. Probable solution: replace the 6-inch edger with an 8-inch by 60-inch single-arbor combination battery and selecting edger. Introduce a canter twin or quad line between the present double-cut and resaw bandmills. This will also involve a high-speed barking-bucking line. It should handle 1,000 to 1,500 logs per shift with piece counts to 14,000.

It must be able to handle logs from 4 to 72 inches in diameter and 8 to 20 feet long. Flexibility is imperative. To illustrate how far small-log piece count is already out of proportion to volume of wood sawn, logs 6 to 16 inches in sugar pine are 32 percent of the pieces but 5-3/4 percent of the volume; in ponderosa pine, 33 percent of the logs but 4 percent of the volume; and in fir, 70 percent of the pieces make 30 percent of the volume.

CHAPTER 5 A WAY TO KEEP YOUR SMALL-LOG SIDE SIMPLE

By Ray C. Swanson

RAY C. SWANSON, president of Swanson Bros. Lumber Co. Inc., Noti, Oregon, has worked in various wood product firms all his adult life. These include a box factory and plywood fabrication plant. He was a head sawyer before entering management where 28 of his 43 years in the industry have been spent. He is also vice president of Swanson-Superior Forest Products, Inc. Always active in the industry, Swanson is a board member and committee chairman of the Oregon Board of Forestry. He is a past president of Western Forest Industries Assn. and serves on the Western Wood Products Assn. board of governors for District 2. Swanson is also board member of the West Coast Lumber Inspection Bureau and of the Timber Operators Council. He is a former chairman of the Oregon State Board of Education.

CHAPTER SUMMARY: The most difficult task is to develop something that is simple, efficient and economic. Pushing a computer button seems simple, but chaos results from a malfunction.

You gain nothing by automating to eliminate six lumber handlers from the sorting chain if you must hire two electricians and four millwrights to make it run. Space limitation and the need for speed in unloading tend to offset the desirable number of sorts which take account, not only of log size and grade, but of best product destination for the log. Swanson Bros.' small-log mill sorts with a Cat 966 front-end loader.

Logs larger than 15 inches are kicked out ahead of the debarker. Logs with minimal deviation are sized by a reducing-ring device in the barker. They are fed to a Stetson-Ross chipper-canter. Reversible shafts with sprockets rotate the log to the proper infeed position. A buckhorn infeed moves it into the chipping heads. A marker indicates on the cant its entry position for the resaw and its modular size for maximum recovery. A knuckleboom loader can stack excess cants for later feed to the resaw. Nominal 2-inch material moves to a two-saw edger. The 4-inch is bypassed to a gang edger. A narrow belt drops edging onto a single-line by-product conveying system. From the time the log enters the mill until lumber is ready to be pulled from the green chain, it is never handled by lift truck or carrier.

All four by-products–chips, sawdust, shavings, and bark–are collected in straight-line conveying systems with small laterals from each machine.

Knowledge leads from the simple to the complex. Wisdom leads from the complex to the simple.

CHAPTER 6 ATTENTION TO DETAIL CAN REDUCE LOST-TIME DOLLARS

By David H. King

DAVID H. KING is president of King, Murphy & Associates Ltd., a wood products consulting firm located at Richmond, British Columbia. He has spent his entire working life in engineering and management in the wood products industry. Before forming his own firm three years ago, he was sawmill manager for Hampton Lumber Co. and for Crestbrook Forest Industries. Earlier, he was manager of sawmill engineering for H. A. Simons International Ltd., consultants. He was plant engineer and production superintendent for MacMillan Bloedel and, earlier, construction superintendent for the predecessor Bloedel, Stewart & Welch. The MacMillan group work spanned 22 years, much of it at Chemainus and Port Alberni, British Columbia. He was involved with redesign and installation of barkers, chippers, log handling systems, headrigs, and edgers.

CHAPTER SUMMARY: Most plants can solve their mechanical inefficiencies, but it is helpful to have an outside source point them out for evaluation. Too often the small problems, which do bear on company profitability, are shunted aside for larger issues. Wasteful practices then become daily routine. Malpractices leading to profit loss mainly concern: (1) machinery shutdown, (2) manpower inefficiency, and (3) product mismanufacture. The latter, of course, represents wasted time and raw material.

Time loss and cause should be tabulated, accounting should supply a figure on the operating cost per minute of the loss, and the supervisor then knows how much money he can justify requesting to rectify it.

One single-line mill estimated downtime at $7.59 per minute. Infeed problems alone were costing $20,000 per year. Losses are from large spaces between logs, logs sliding back, and logs cradling in the trough. Run the chain more slowly but add chairs.

Plastic wear strips reduce friction, reduce horsepower needs, and virtually eliminate wear. Drive chain, elongated by wear, can be detected by a floating idler which activates a cam switch, telling the operator lengths should be removed to prevent the chain jumping off the sprocket.

Debarker abrader arms should get the same attention as saws. Log centering and maintaining correct self-opening curvature of the arms are critical. A fluted roll case and a pair of fixed chop saws will increase cutoff station production where the same length log is wanted repeatedly.

There is an analysis of proper feed and saw speeds. Considerable concentration is on materials handling equipment.

CHAPTER 7 CONTROL LOGIC FUNCTIONS INEXPENSIVELY AND PRECISELY

By Robert P. Lewis

ROBERT P. LEWIS is engineering manager of Lewis Electronics, Inc., which he established at Hillsboro, Oregon, in 1971. His background in electronics and computers began when he was an instructor in the field at the Naval Air Technical Training Center at Memphis, Tennessee, after a basic U.S. Navy education in airborne communications and radar. He spent two years on the military SAGE project (radar warning and BOMARK missiles). Lewis spent five years in Alaska as a systems engineer for the Federal Aeronautics Administration. He worked with Federal Electric Corp. and as a staff engineer on aerospace subsystems. Lewis Electronics has developed many sensing and control systems for the lumber and plywood industry.

CHAPTER SUMMARY: Cost of sawmill control and recording systems is dropping rapidly with recent technical developments in integrated circuits. A simple limit switch which activates a solenoid is a "logic" system, or embryonic sensor-computer-control. Systems can be simple. And far more sawmill functions can therefore be automatically controlled at a financial saving than the industry has understood to be possible. Similar cost reductions have taken place in hand-held calculators.

Universal logic elements for computers vary in power depending upon design for (1) type of instructions it can accept, (2) word size, (3) speed of its cycle in obtaining information or performing a task, and (4) how big a problem it can handle–including the size of its memory system.

Once a mill has a computer and an extra set of cards, many functions in the mill can be hooked into it. Cycle time for a computer to sense or scan a point in the production line is exceedingly brief. Most systems can handle many more tasks than they are set to perform.

Many operating problems can be easily handled. For instance, electrical contacts, when closing, tend to chatter. If the computer scans at the moment the contacts have bounced open, it will misread the closure. But the computer can be instructed to take its reading a fraction of a second later, getting the right information.

Numerous output devices can be used–relays, solenoids, motor controllers, numeric displays, and printers among them. Each must have its program stored so the computer can instruct it what to do.

Mill computers should have as much built-in test equipment as possible so that breakdowns and malfunctions can be found quickly with minimum downtime and/or mill visits from outside computer maintenance men.

CHAPTER 8 RAISE SIDEBOARD RECOVERY WITH COMPUTERIZED EDGER

By Chris Vadnais

CHRIS VADNAIS is sales manager for Saab-Totem, Inc., a subsidiary of Saab-Scania of Sweden and Totem Equipment Co. of Seattle, Washington. The subsidiary was formed in 1975 to manufacture, market, and service the newly introduced edger optimizer to the forest products industry in the Pacific Northwest. Born at Pendleton, Oregon, Vadnais attended Washington State University, graduating with a bachelor's degree in economics. He spent two months in Europe studying the operation and performance of the new edger before its introduction here.

CHAPTER SUMMARY: Board edging has been little improved. The Saab interests of Sweden began to develop an automatic, computerized edging system in 1973. The resulting edger optimizer was presented to the Pacific Northwest industry by Totem Equipment Co. in November 1974.

The edger optimizer puts 92 percent of sideboard volume into lumber, compared with a range of 70 to 85 percent for manual edger operators. Machine functions include: (1) feeder with overturn and reject options, (2) thickness sensor, (3) face (and length) scanner, and (4) positioning device to optimize saw or chip edging.

The edger optimizer takes sideboards from a quad or a Chip-N-Saw. It is preceded by an automatic unscrambler which assures single board feed. There the operator can feed, overturn to place wane face up, or reject. The operator tells the computer the highest grade the board can make.

Bright light scans for width and detects wane. The board is scanned every 4 inches and the computer decides upon a reference line for the edger. Inputs the computer receives to evaluate volume and price relationship include: (1) grading rules on allowable wane, (2) market relationships among sizes and grades, (3) thickness, (4) face and length information on the piece to be edged, and (5) grade potential as plugged in by the edgerman.

The board is transferred to an infeed chain, skewered, centered, and edged. Chipheads and saws are movable.

Recovering the difference between 85 percent and 92 percent of the cant volume as lumber instead of chips (minus extra downstream production costs) should increase recovery $182,000 yearly for a small-log mill producing 60 million-board feet and getting one-third of its volume in sideboards.

CHAPTER 9 POINTS IN SELECTING A WASTEWOOD CHIPPER

By Fred Veuger

FRED VEUGER is vice president for marketing at CAE Machinery Ltd., Vancouver, British Columbia. Born in the Netherlands, he has lived in Canada since 1959. He holds degrees in mechanical engineering and business administration from the University of California and the University of British Columbia. Veuger joined CAE (then Canadian Sumner Iron Works) in 1960, becoming assistant chief engineer. In 1967 he was hired by the Council of the Forest Industries of British Columbia and posted in Europe to develop lumber and plywood markets. He joined Tahsis Co. Ltd. in 1971 as chief of engineering services at its sawmill and pulpmill on Vancouver Island (annual capacity 180 million board feet). He returned to CAE in 1972 in his present post. Veuger is a member of the Assn. of Professional Engineers of British Columbia and president of Canmillex Sawmill Exports Ltd., a consortium of six major sawmill machinery manufacturers exporting overseas. The talk, prepared by Veuger, was actually given by Jim Hatton, the company's sales manager, when a travel schedule conflict developed and prevented Veuger from speaking at the Clinic.

CHAPTER SUMMARY: Wastewood or mill refuse chipping presents different problems from whole-log or headrig chipping. Many of those problems result from lack of forethought in original machine selection. Yet the chipper is an important piece of machinery—a chipper shutdown will soon lead to a shutdown at the headrig.

Types of refuse to be chipped should be studied before a chipper is selected. Long material should go to a horizontal-feed chipper. Slashing it only reduces chip recovery.

Short blocks are better fed to a gravity machine. Vibrating conveyors have lessened the problem of jamups in the chipper spout, but unfortunately their feed rate is low.

A top-discharge chipper wastes energy in blowing or throwing the chip up, requires more ducting, and is noisy. If used, care should be taken to prevent throwing the chip against the duct.

Generally, the heaviest chipper for a given spout size will have a number of advantages. It will produce the best chips, require the least maintenance, and operate the longest.

Good chip quality requires least clearance between knife and anvil. To achieve this without running the risk of the knife striking the anvil, the stability of a tapered disc, a heavy arbor, and bearing support from both sides is desirable.

Wood needs a stable support against the anvils and, since mill refuse usually has sawn faces at right angles, the anvils should always be at right angles to each other.

Protection from tramp metal is vital and a metal detector is the best device. When metal or rock gets through, action must be immediate, as the disc is sweeping through nearly 100 knife cuts per second. CAE pivots the side angle and holds it with shear bolts. When pressure is excessive, the shear bolts let go so metal or rock is discharged through an opening in the spout's side.

Carefully-chosen and well-maintained chippers can successfully turn waste and refuse into profit.

CHAPTER 10 IMPROVING DRIVE EFFICIENCY WITH SYNTHETIC FLAT BELTS

By Greg Baird

GREG BAIRD is the Northwest operations manager for Habasit Belting Inc., located in Seattle, Washington. Born in Spokane, he obtained his bachelor's degree in mechanical engineering from Gonzaga University, also in Spokane, in 1955. Baird worked extensively in welding engineering and lubrication engineering as well as as a metallurgist. He operated heavy equipment in logging and construction before heading his own sales and engineering company. He worked in the ordnance division of FMC Corp., for Smith-Berger Manufacturing Co., and for Stetson-Ross Machine Co. In full-time sales for Habasit for the past several years, he covers the Pacific Northwest and northern California. He belongs to the Assn. of Operative Millers, and to the Technical Assn. of the Pulp & Paper Industry (TAPPI).

CHAPTER SUMMARY: All belts used for power transmission are governed by these principles:

1. They require a high coefficient of friction.
2. Elasticity of the material must be balanced through the length of the belt.
3. It must be lightweight.
4. It must have little induced vibration.

These factors must be optimized to deliver the most power for the least cost. For instance, the coefficient of friction should remain the same regardless of changes in the speed at which the belt is run. The coefficient changes with V-belts as the belt rubs in and out of the groove.

Only during the past 50 years has belting been made from other than natural materials such as leather or cotton. Natural materials were unpredictable and uneven in wear, elasticity, friction, and other characteristics.

A rigid material transmits shock heavily back to the motor. Belt material should be elastic enough to absorb most shock.

V-belts are not efficient above 5,500 or 6,000 feet per minute, whereas a flat belt remains so and can pull 120 horsepower per inch of width at 10,000 feet per minute.

Vibration destroys bearings rapidly. Most vibration is caused by belt splices. A misaligned flat belt will not run at all, but a misaligned V-belt causes vibration. Field comparisons are shown for all-synthetic flat belts, combination leather-nylon belts, and V-belts in driving bandmills, gang edgers, and chippers. As an edger drive, for instance, the leather-nylon belt, with a lower coefficient of friction, creates a higher bearing load. Diameter and location of pulleys greatly affect results.

CHAPTER 11 BUILDING A PRACTICAL QUALITY CONTROL PROGRAM

By Dean W. Huber, Geroge H. Sisterhenm, Robert L. Mardock, James L. Russell, Robert W. Nix & Frank John Banducci, Jr.

DEAN W. HUBER has joined the U.S. Forest Service's program to analyze sawmill efficiency, having left his private consulting firm at Lake Oswego, Oregon. The move to the Forest Service's Forest Products Utilization program at San Francisco, California, has continued his first-hand contact with mills throughout the West. Huber began with summer employment as a sorter operator and green chain puller for Weyerhaeuser Co., Longview, Washington. He obtained a master's degree in wood utilization from the University of Idaho in 1972. His father is head sawyer at a mill in Washington. Huber suggested the topic and assembled the speakers for a panel on the practical side of quality control.

GEORGE H. SISTERHENM is project manager for the Forest Products Group for Fibreboard Corp., San Francisco, California. Much of his work concerns the sawmill at Standard, California, however. Born at Durham, North Carolina, he obtained a bachelor of science degree in forestry from the University of Massachusetts and followed it with a master of forestry degree in wood technology from Yale University. Sisterhenm spent eight years as an operating plant manager for a plywood mill before moving into business planning and development. He is now on a special project assignment for Fibreboard. He is on the marketing and promotion and the technical advisory committees to the American Plywood Assn.

ROBERT L. MARDOCK is the operations manager for Harwood Products Co.'s 25-million-board-foot-per-year redwood and Douglas fir sawmill at Branscomb, California. Born at Oroville, California, he served as plant manager for Crosbie-Bamert, Inc. before joining Harwood Products. He started in sawmill work with the trade of electrician but his interest in machinery design and operation has carried him into systems analysis. With Dean Huber, he recently presented a program on increased recovery from edger operation through machinery modifications in setting saws into the correct position. The Forest Products Research Society's northern California chapter meeting at San Francisco was the setting. Mardock is a trustee of the Willits unified school district and a past president of the Laytonville Lions. He is a race car driver in his spare time.

JAMES L. RUSSELL is the quality control manager for Idaho Forest Industries' sawmill at Coeur d'Alene, Idaho, but lives at nearby Post Falls. Born at Pomeroy, Washington, in the "Inland Empire" area where he still lives, Russell spent 24 years with a Potlatch, Inc., sawmill–10 of them as a dry kiln foreman–before joining Idaho Forest Industries. He joined IFI in the spring of 1971 in the same position. He then moved through shipping and planer foreman jobs before becoming production manager of an IFI plant for about two years. The experience of directing the actual operation of so many departments prepared him for the quality control manager responsibility he has held for the past two years.

ROBERT W. NIX came to his job as utilization supervisor at Diamond International Corp.'s Coeur d'Alene, Idaho, mill via the University of Idaho. There, he majored in wood utilization and minored in business, receiving a forestry-wood utilization degree in 1971. En route, he performed a summer and fall project for Boise Cascade Corp. on drying Philippine mahogany and ponderosa pine peeler core studs. The 1970 summer went into initial testing of the Plystran concept for Potlatch Corp. In 1971 it was a job for McFarland Pole Co. on use of chromated copper arsenate on Douglas fir poles. He joined Diamond that year in quality control and has been a temporary sawmill foreman and utilization supervisor. Nix belongs to the Forest Products Research Society and the Inland Forest Resource Council and is interested in seeing the industry viewpoint conveyed to the public.

FRANK JOHN BANDUCCI, JR. is quality control supervisor for Roseburg Lumber Co.'s Flakeboard operation at Dillard, Oregon. Born at Bakersfield, California, he finished high school at Pleasant Hill, California in 1958, then ran the family ranch for several years, coming to the Roseburg plant in 1965. He supplemented hard work and experience with a correspondence course from International Correspondence Schools. He started as a quality control checker, working all areas of the Flakeboard plant–the line, cut-up, shipping, sanding, and so forth. In addition to production qualtiy control in the plant, he is responsible for all raw material quality and handling, for safety, and for the product development program. He belongs to the Forest Products Research Society and to the American Forestry Assn.

CHAPTER SUMMARY: This chapter on practical quality control records a panel of speakers who divided its aspects–management objectives, application to machinery, and motivation of people are three. It was arranged by Dean Huber, who works on the Sawmill Improvement program conducted by the U.S. Forest Service through its Division of State & Private Forestry.

He emphasizes that the mill manager must control performance at his mill, not be controlled by it. He must be a thermostat to make things happen, not a thermometer to reflect them.

George Sisterhenm recounts the need for management to set objectives, to produce to them, and to erect a system of accountability. He notes the rise of consumer quality expectations and the corresponding increase in industry's technical ability to control quality. He notes the need to involve plant personnel from the beginning.

Bob Mardock relates the improvement in performance which thoughtful analysis of machine functions can bring about. Careful machine monitoring is the basis both for quality control and for good recovery. A simple rack and pinion, for instance, moves the bank of edger saws to and from the linebar, improving line-up of the No. 1 saw into the wane edge–improving recovery.

Jim Russell describes Idaho Forest Industry's efforts to involve people in its quality control programs. Right personnel selection to transmit quality objectives is imperative. Quality control people must have mill experience, the respect of line supervisors, and the ability to communicate objectives. A manual has been developed to cover each job at IFI.

Bob Nix explores the improvements that can be had from good quality control in just two areas: log storage (where he discusses sprinkling) and the reduction of trim loss and degrade at the dry kiln and planer. Volume and grade loss and customer complaints about end check in lumber from unsprinkled logs are cited.

John Banducci reviews the quality control program at Roseburg Lumber Co.'s particleboard plant at Dillard, Oregon, the world's largest. Quality control begins with a sample lab test of each incoming truckload of raw material. He shows the extreme precision of the resin mix, the importance of particle size and moisture content of the furnish, the careful control of the mat. Product uniformity to 0.005 inch and care in sanding and packaging are described. Past performance reports are studied to spot future tendencies.

CHAPTER 12 COMPRESSED AIR–WITHOUT OIL-COOLANT WORRIES

By John G. Greenwood

JOHN G. GREENWOOD is vice president for the industrial division of Sullair Corp. Born at Hartford, Connecticut, he graduated in industrial engineering from Lehigh University, Bethlehem, Pennsylvania, and has been associated with the air compressor business for 37 years. He started as a serviceman and worked through engineering and sales positions for three compressor manufacturers. He now travels throughout the world. Although the Sullair plant is at Michigan City, Indiana, Greenwood has long lived at Essex, Connecticut, where he has the largest private collection of sailing ship figureheads in the world, some of them from 19th century British ships which were forced to get Douglas fir from the Pacific Northwest when post-Revolution trade restrictions cut off the traditional source of East Coast pine.

CHAPTER SUMMARY: The rotary screw compressor, a great advance, was invented by the Swedes in the early days of the jet engine. It was lightweight and eliminated surge or pumping.

Air compression should be simple but is made difficult by the generation of heat. A single-state piston compressor produces temperatures above 400 degrees Fahrenheit. Vane compressors were popular in the Northwest with low electric costs but are inefficient.

The rotary-screw compressor has only two working parts–helical rotors that rotate together. Power is in the male rotor. A dry rotary requires closer tolerances and produces high temperatures. The flooded screw compressor provided coolant and began with mineral oil, injected with the air, as coolant-lubricant. It also provided a seal between rotating parts and cylinder border. There is no wear between rotors, and lubrication need is minimal. In oil-cooled screw compressors, air and oil reach only 170 to 180 degrees. That temperature is sufficient to oxidize mineral oil, however, creating grit. It has taken a long while for mechanics on air-cooled engines to realize grime is not introduced through dirt in the air intake, but is produced by chemical changes in the oil, particularly under high temperatures.

A recent advance has introduced a new "oil" which does not oxidize under heat. Sullair uses this on flooded screw compressors producing 85 to 3000 cubic feet per minute of 100-pound air. The base for the coolant, 24KT, is silicone oil or liquid silicone, derived from sand. It shows little change in viscosity with temperature change. It is about half as inflammable as mineral oil. It is not toxic. It is a poor lubricant, so the manufacturer has developed an additive. It never needs changing during the life of the machine.

CHAPTER 13 OPENING THE FLOW FROM BULK STORAGE CONTAINERS

By Rick Bickers

RICK BICKERS is sales engineer for Stephenson Industrial Supply, a Portland, Oregon, firm. Previously with Martin Engineering Co. as manager of industrial applications, he has also been associated with Shalco Systems (Acme-Cleveland Corp.) and with Demmler Manufacturing, both in the foundry equipment field. Born in Kewanee, Illinois, Bickers attended the Morrison Drafting Institute. He is a member of the American Foundrymans Society and has published technical papers on a number of subjects.

CHAPTER SUMMARY: The Big Blaster is a new concept in keeping the flow open in bulk storage containers. Industrial vibrators, however, are needed for some applications.

The Big Blaster has only one moving part: a polyurethane piston which is nonsparking. It works off compressed air from 60 to 125 pounds per square inch. Discharge takes 4 milliseconds. Control is a three-way operating valve, normally closed. Operation can be manual or electric. When open, outside air seats the piston and enters the pressure vessel.

Release force compacts the material in its path until resistance is overcome. The Blaster is normally installed pointing toward the discharge end.

The wood products industry has all of the problems of bridging, arching, and ratholing, the last drastically reducing the effective storage area available. Clearing is expensive, either in tools or man-hours. And the problem soon reappears. A mass of compacted material may fall, plugging the discharge.

The Blaster is trouble-free. Units cycled 350,000 times show virtually no wear. A unit cycled two shifts, six days a week would complete less than 20,000 cycles per year.

More than one unit is sometimes needed in a container. Shavings, hogged fuel, and chips require a larger Blaster than sawdust, as they contain more air holes which the Blaster air must fill before applying its pressure.

Electric or pneumatic vibrators are often used. The electrics require much maintenance and are self-destructive. Noise levels may be 110 to 125 decibels. They sometimes pack a bridge tighter instead of removing it.

CHAPTER 14 DRY KILN DESIGN AND WOODWASTE RECOVERY

By Jack L. Bachrich

JACK L. BACHRICH, the president of Intercan Mill & Kiln Co. Ltd., North Vancouver, British Columbia, was born in Austria. He attended technical high school and technical university in Vienna before moving to England. There he operated Bachrich Kiln Co. until 1958, when he emigrated to British Columbia. He managed the Bachrich Dry Kiln departments of Galbraithe Sulley Ltd. and of Forano Ltd. until 1975, when he set out with his present firm. His work on dry kiln design, construction, and operation has been published in many professional and trade periodicals and he is at work on a book in this field for Forano Ltd. now. Many mills have adopted his ideas on using the energy from mill waste to dry their wood.

CHAPTER SUMMARY: Fuel-cost increases on the West Coast have been immense. This chapter examines the implications for dry kiln activity, including: (1) fuel costs, (2) heat requirements for kilns, (3) cost of heat from fossil fuels, and (4) use of woodwaste as fuel for the middle-size sawmill.

Per million Btus of heat energy contained, Bachrich rates the cost of natural gas at about $1 in Vancouver and in northern British Columbia and at $2 in Oregon. He rates butane at $2.80 to $3 in northern British Columbia and at $3 in Oregon. He rates Bunker C oil at $2.20 in Vancouver and at $2.14 in Oregon. In contrast, he points out that dry shavings cost only $0.36 cents in Oregon (based on a per-ton price of $6.50). Where it is burned as refuse, with no energy recovery, wood refuse has a negative value. Shavings should be worth $36 a ton against natural gas and $54 against butane.

Roughly two pounds of steam are required to evaporate one pound of water from green lumber. Gas-heated kilns are less efficient than steam-heated kilns because products of combustion must be bled off from the former, as they must from all direct-fired hot-air kilns where there is no heat exchanger.

A peculiarity of Canadian law requires the presence of an operating engineer with high-pressure steam kilns. Hot-water systems of equal heat value do not require a full-time operator so wage costs can be saved in Canada by a hot-water system. Boilers should have a somewhat greater capacity than the two pounds steam/one pound water evaporation formula would appear to require as the extra capacity will lengthen boiler life.

A woodwaste fuel storage bin capacity of 72 hours should be considered, but often low-cost cribs in close proximity can permit a front-end loader to feed the bin, with less capital outlay.

CHAPTER 15 OPTIMUM USE OF YOUR TOP RESOURCE—PEOPLE

By Jay Gruenfeld

JAY GRUENFELD is vice president for lands and forestry of Potlatch Corp., at Lewiston, Idaho. Since mid-1973 he has been responsible for 1.3 million acres of diverse timberlands in Idaho, Arkansas, and Minnesota. He spent 18 years with Weyerhaeuser Co. in western Washington, moving from choker setter through forester, manager of a 70,000-acre industrial tree farm, assistant land supervisor, logging superintendent, and manager of timber and log sales. For 3½ years he was raw material manager for Brooks-Scanlon, at Bend, Oregon. Gruenfeld earned his bachelor's degree and a master's in forestry at Colorado State University. A Fulbright scholar, he received a diploma from Oxford University in humanities and in forestry. He is vice chairman of the Forest Practices & Land Use Action committee of the

American Paper Institute and the National Forest Products Assn. He is active in the Society of American Foresters and other industry groups.

CHAPTER SUMMARY: Great strides in sawmill-plywood plant technology are coming. But this is not the area where the greatest improvement in resource use can be achieved. It can be achieved in better use of people.

A top logger told me, "I'll give you your choice of equipment and you give me my choice of the men and I'll outlog you every time." He was right. Often plywood mills with inferior equipment outproduce those with better machinery because the crew is better trained and better motivated.

Most problems involving people involve a failure of communication.

Potlatch emphasizes the importance of two-way communication with employees. The firm publishes a booklet, for distribution to all employees, discussing the company's philosophy, objectives, and values. New employees have a much better opportunity to understand company policy with it. It is apparently the only such booklet in the forest products industry.

Potlatch's Community Council, an outside group, advised the company to communicate with employees on a two-way basis. This would arm employees to tell their circle of acquaintances about the importance of the industry and the impact such problems as the Monongahela decision could have on jobs and on the community's economy. People act according to what they believe to be the facts—not necessarily according to what the facts really are.

They support what they have helped create. Involve employees, for instance, in the writing of a company booklet.

People are the top resource.

CHAPTER 16 HUMAN FACTORS IN ENGINEERING DESIGN

By Joe G. Sublett

JOE G. SUBLETT is manager of industrial design and human factors at Weyerhaeuser Co., Tacoma, Washington. He is concerned with expanding knowledge about people's capabilities and reactions and applying that to the industrial worksite. Changes in sawmill machine operator work stations are an example of this approach. Born at Huntsville, Alabama, he majored in psychology and obtained a bachelor of industrial design degree from Auburn University. He went on to a master of science in systems management from the University of Southern California. Before joining Weyerhaeuser, Sublett was manager of crew systems design for Aerojet-General Corp. and was a project engineer for flight deck design at Boeing Co., Seattle, Washington. He is active in the Technical Assn. of the Pulp & Paper Industry (TAPPI), the Human Factors Society, and the Industrial Designers Society of America.

CHAPTER SUMMARY: Man has been expected to fit into his occupational environment. This has wrought stress, poor performance, and even physical harm to the worker. High turnover, absenteeism and union dissatisfaction have been the results.

Development of accurate data about people did not parallel information about machine processes during the Industrial Revolution. But the more sophisticated the equipment, the more vital the operator's role becomes. To extend man's capability it is necessary to design from the man out. Considering man's capabilities and limits leads to the creation of the best man-machine system.

Man's senses make him an excellent information receiver, with input peaking at about 10 successive bits per second. Confusing or overlapping inputs should be avoided.

Man is an exceptional evaluative computer. He is the only computer that can solve problems of logical induction. But man's outputs—such as operating control buttons and levers—are relatively slow and low-powered. He is, however, flexible.

To use him to best advantage, information is being brought in from many fields to create the two newly emerging fields of industrial design and human factors engineering.

The scope of industrial design/human factors design includes: (1) systems/operational analysis, (2) operator environment design, (3) work station design, (4) operational safety, and (5) operator training.

In a mill environment, the positioning of levers, readouts, seat, and so forth, should be done with the individual operator's sight line, range of motion, and other physical characteristics, in mind. There is no average man. Color coding and movement relationships should be those expected. Red is stop; green is go. Clockwise increases; counterclockwise decreases. Avoid both excessive stress and boredom.

The cost of applying these factors should be considered against the potential gain.

Human performance should be considered as part of the work system—not isolated.

Much information has been developed lately about human performance. It should be applied to man-machine system design.

CHAPTER 17 THE IMPACT OF LAND-USE PLANNING

By V. M. "Whitey" Howard & William Lesh

V. M. "WHITEY" HOWARD is both one of the most thoughtful and one of the most vocal critics of the reduction of the West's commercial timberland base. He is vice president and general manager of Seneca Sawmill Co., at Eugene, Oregon. Born in Portland, he attended Vanport Extension Center and Oregon State University before graduating in forestry management from the University of Washington. He worked for the U.S. Forest Service on the Mt. Hood and other national forests before working with the former Portland Manufacturing Co. He was staff director for Western Wood Fibre, Inc., a chip marketing vehicle for a group of independent mills, and is now a vice president there. He is also a vice president of the Western Environmental Trade Assn. (WETA) and belongs to the Society of American Foresters.

WILLIAM LESH is Publishers Paper Co.'s assistant director of public relations. As such, he has become vitally involved in land-use decisions throughout the Douglas fir region. A native Oregonian, he graduated from the University of Oregon with a degree in economics in 1967. He served on the staff of the Multnomah County (Portland area) Republican Central Committee and for Ed Nichols Associates before joining Associated Oregon Industries (AOI) as member relations director. He is now chairman of the land-use committees for AOI and for the Western Environmental Trade Assn. (WETA). He serves on the Board of Governors of the Portland City Club and is its program chairman. Lesh was recently a candidate for the Oregon legislature. He lives his advice to other businessmen about land-use problems: "Get involved."

CHAPTER SUMMARY: Part one of this chapter is Whitey Howard's discussion of the impact of land-use planning on public timberlands and its adverse effect upon the western regional and the national economy.

He calls it "arbitration between the takee and the takor—or who gets the benefits and who pays the costs." Public lands belong to all the people and should serve all needs. Forest Service land-use plans in the West show a reduction in fiber production about 20 percent below current harvest levels, meaning lost jobs, higher taxes, and more wood imports to meet national needs. In the Pacific Northwest, Forest Service reductions would cost 22,100 jobs, reduce payrolls $270 million yearly, and cost counties $50 million in in-lieu tax payments. The Administration and Congress have set no quality nor quantity goals for the U.S. Forest Service so planning becomes a political football game between the biological ability of the forest and the projected needs of people.

When a forest land-use plan has been adopted, it is "set in concrete" for 10 years. Forest Service plans contain innumerable nonproductive restrictions. On some forests–notably the Willamette, which is the nation's most productive–industry has proposed more productive alternatives and these are obtaining good public support.

Bill Lesh urges the industry to get involved in land-use planning decisions lest they be made by the zealots, the bureaucrats, and the uninformed–to the detriment of the economy.

Management cannot simply assign land-use planning problems to a staff man, then ignore them. First understand the "bureaucratese" and "legalese" in which those who wish to plan and control land-use decisions talk. Understand the structure of the governmental agencies and the controlling court decisions. Planners paint colors on maps–dictating what people can or cannot do–without regard to legal boundaries or economic realities. Know what they are doing to you!

CHAPTER 18 THE NATIONAL FORESTS: SOME NEGLECTED COSTS

By Barney Dowdle

BARNEY DOWDLE has challenged traditional U.S. Forest Service timber management ideas for some time. Recently, his theme that the national forests are an asset which should pay the public a return on their value has received increasingly widespread favorable discussion. Dr. Dowdle is a professor of economics and forest resources at the University of Washington's College of Forest Resources, Seattle. He received a bachelor's degree from that university in 1957 and his master's and doctor's from Yale. Dowdle was a logger from 1946 to 1951 and worked for the U.S. Forest Service from 1959 to 1962. A Phi Beta Kappa, he belongs to the American Economic Assn. and similar professional groups. He recently organized an economists' symposium on "The Economics of High-Yield Forestry" at Seattle, which was severely critical of Forest Service timber management.

CHAPTER SUMMARY: U.S. Forest Service management of the national forests results in a net loss to their owners—the American people. This is because the Forest Service refuses to recognize the value of the asset—land and timber—and its obligation to obtain a return on the investment. Proper management could change the national forests from a net liability to a net asset to the public. Present management is "irrational, inefficient, and a burden to society." Worse, it redistributes income from the poor to the rich.

The owner of any asset has the option of selling it and investing his money at interest if the interest exceeds the income from the active management of the asset. Cash flow from the national forests, however, represents about 1.1 percent of their value. Deduction of carrying costs of the land and timber inventory would throw them into a loss position. In Oregon and Washington alone, the loss would be about $1 billion yearly. This is the cost of permitting the Forest Service to continue its sustained-yield nondeclining evenflow policy.

Exxon, with assets of $31 billion, netted $3.1 billion in 1974. The Forest Service, with assets valued at $42 billion, lost $2 million that year, a study by Dr. Marion Clawson, economist and former head of the Bureau of Land Management, estimates.

The Forest Service believes the price system will not work in forest management. It works for the South, however, where forests are privately owned. The South harvested more than 300 billion board feet of sawtimber between 1950 and 1970. Inventory in 1950 was 357 billion board feet and, in 1970, 438 billion—up 81 billion despite cutting.

Forest Service policy is held over from feudalism, when it became necessary to limit access to community-held woodlands for fear of overcutting. Establishment of private, transferable rights created a market system and the overcutting threat vanished.

1

Dry-land log handling: dump to infeed deck

Scott Corbett, Jr., President, Howard-Cooper Equipment Co., Portland, Oregon

The evolution of dry-land log handling was accelerated by the rising cost of the tree and of labor. The secondary rising cost and limited availability of land was incidental to those factors. The increased cost of the tree necessitated an appraisal of the best return in processing it—pulp, plywood, lumber, stud, or whatever. And, with markets opening up for utilization of sawmill residuals, the necessity to sort to species to enhance chip sales increased.

The cost of the man on the pond, particularly when governed by the boom man's union, increased the cost of separating logs by species and then by diameter and then by grade. Increases in mill productivity meant more raw material could be put through the same source and location. This necessitated a greater reserve or a deck area; and the complexity of processing smaller trees, and thus more pieces, quickly enhanced the value of dry-land mill yards.

A look at the history

The early-day Pettibones with the Lature forward extension patent in the 10,000-, 15,000-, and 30,000-pound capacity range were the forerunners of dry-land log handling with forklifts. The first production-model large forklift was built by R. G. LeTourneau. I personally had the pleasure of working with him in the development of that unit. But the potential of the concept of dry-land log handling was not known to any of us then. The forklift was designed to release the truck from the scaling station more quickly.

I also had the pleasure of selling the first Raygo-Wagner Lumberjack. So I have been intimately connected, for good or bad, with the evolution of the dry-land mill yard from its beginnings. Then, the concept was to release the trucks faster, to hold the truckload as a unit until it was scaled at a dry-land sort, to transport it to a deck which could be built upward to reduce total acreage, then to move it from deck to infeed. Crane and heelboom development—upgraded swing and lift mechanisms, heelbooms, hydraulic grapples, and so forth—paralleled lift truck development in this basic area of log handling.

Today the dry-land mill yard is undergoing revolutionary change. I say this because the factors that prompted early dry-land log handling—(1) concentration of facilities, and thus increased volume in one location; (2) high log costs; and (3) the need to make a choice of what to do with a log or a portion thereof—are even more pressing today and are stimulating interest in what can be done in the mill yard and in the specialized equipment now available or being developed.

The main point I will stress throughout is this: The mill yard is a "profit center" unto itself.

Close attention to mill yard can offer dividends

My experience has shown that a number of operations have not yet accepted the idea that the mill yard is an operational area which contributes to the profitability of the total operation. In the early days, I could understand this. Then, the logger logged. He built the roads out in the woods; he handled reforestation; he cared for the logging trucks and the skidding equipment; he brought the log to the pond or to the mill yard. And then he forgot it.

The sawmill man's responsibility extended from the infeed through the sawmill—optimizing return on the logs which came to him. Who took care of the equipment involved in between or thought of the return on an individual log? I will be frank. Hardly anyone. Who handled the maintenance of the mill yard? No one. Who paid attention to the benefits that could be received by sorting? Really, no one. The mill yard was a no-man's land. The sawmill man did not care about it. All he was really concerned about was to get that log into his mill and volume—not necessarily "piece profitability."

With the increased return available through specialized handling, many now look at the mill yard and recognize it as a specialized area. The sawmill man is not totally equipped to handle it; the logger is not totally equipped to handle it. But with the equipment available today, it has become important. It represents great potential for savings, as well as for increased production, to the total operation.

For a machinery peddler to address a group of potential customers without trying to sell something specific is very difficult—and it is particularly difficult for me. For a machinery peddler to attempt to present a functional operational technology to an end user whose business is making money from the total processing of wood fiber would be not only difficult, but ridiculous. Therefore, all I will try to do is point out systems that are available and in use today handling logs in the mill yard.

Mill yards representing various stages of advancing technology can be found throughout the industry. Needs are different; times are different. It is important to remember that what might look archaic today was extremely foresighted not long ago. To change completely to total mechanical sorting and scaling and tree "merchandising" overnight might not even start to make economic sense in many cases, but steps are being taken. I am not an overall expert, nor is any one person. A great many people are available to talk with about what is being done or planned. You are developing specialists in this area within your own companies. It is an exciting opportunity and needs continued appraisal.

Before answering the question "how?", we need to answer the question "why do we want to?". The answer to the second question must be based on the individual operator's problems—type of timber, location, facility, price per log. Each one of you will have a different reason "why" you want to do something to the logs in

Top: Wagner lift truck unloads full truckload of logs in mill yard. Capacity of the machine is 90,000 pounds. *Bottom:* Four-wheel-drive Hough 560 lift trucks transport barked logs in the mill yard.

your yard. Allowing for individual problems, however, the first thing that generally happens is the necessity to unload the logs from trucks or rail cars.

Options available in the dump area

A great number of machines are available to the consumer to unload trucks: winches, cranes, grapples, and lift trucks. In lift truck capacities of 70,000 pounds and greater, the Raygo-Wagner Lumberjack and the R. G. LeTourneau stacker dominate the field. Other devices, such as the Kenworth Dart, have this lift capacity; and the largest Caterpillars, Houghs, Terexes, Clarks and some other units are at the minimum truckload weight capacity. But if you are looking for a 70,000-pound-or-greater-capacity lift, the two leaders are Raygo-Wagner Lumberjack and R. G. LeTourneau; the third available unit is the Kenworth Dart.

Forklift units are built up to 200,000-pound capacities. These are highly specialized due to the very limited demand. Many units are built by all three manufacturers in the 120,000-pound capacity. In the 120,000-, 80,000-, and 90,000-pound units, there are many examples of equipment for rail unloading and transferring at satellite yards. For truck and rail unloading, the following models are common choices: Raygo-Wagner L-90 and L-130 (a recent development with a 130,000-pound capacity); 120 Dart Kenworth; LeTourneau CS-803; and Terex 72-71.

In addition to lift trucks, many cranes and grapples are in use. Bear in mind that the surface conditions of a yard could make a four-wheel-drive unit necessary. In 90 percent of mill yards, any of these large units will perform. Units—such as built by Caterpillar, Hough, Terex, Clark, and others—with the oscillating center pin and four-wheel drive have some advantages in turning radius and traction. It is necessary to educate both the equipment manufacturer and the sawmill man that the surface area has a direct bearing on the speed of log handling and the equipment maintenance cost. Both must learn that lift centers and strains are different from those in the construction equipment field. Some examples of grapple and crane handling equipment include: Soderhamn Bush 45-ton grapple; Caterpillar 980B; Hough 560; and Bucyrus-Erie 20HL.

Both rubber-tired (*top*) and tracked (*bottom*) hydraulic grapple loaders are used for sorting in the mill yard. Equipment at top is a Drott 80. An FMC Link-Belt is shown in the lower photograph.

Two options for the unloading operation. *Top:* Crane unloading with slings in a pulpwood operation. *Bottom:* Grapple unloading with a 45-ton Soderhamn-Bush log boom.

Top: Crown Zellerbach has an elevated scaling table at its Lewis & Clark operation. Hydraulic loader with grapple handles logs. *Bottom:* At SD&S, Stevenson, Washington, ground scaling chains are used. Today's log prices make dry-land scaling imperative.

Keep in mind also the safety value of encompassing the load with hold-downs; the stake clearance in lifting heights; the necessity in a yard to provide for reloading the trailers; the need to design the yard with an unloading area so the flow of traffic is unimpeded; and finally, the proximity of the unloading area to the next step in handling.

Approaches to dry-land scaling vary

After unloading, the next step usually is scaling. Here, again, a number of methods are available. Scaling tables or ground chains, as devised by Crown Zellerbach's Lewis & Clark operation (Clatsop County, Oregon) and by SD&S at Stevenson, Washington, illustrate one approach to scaling. Other approaches are scaling bays against stationary logs or through-traffic scaling bays. It is interesting to reflect on how difficult it was in the early days to

Tracked heelboom grapple loader sorts and loads truck at woods landing.

convince mills of the virtues of ground scaling when today it is regarded as an absolute necessity. Thinking back on the 8 to 12 percent defect missed by water scaling, it is almost inconceivable with today's log prices *not* to scale on dry land.

Sorting developments include upgrading

The next step in log handling is where most new advances are taking place. As long as you have handled the log (from the truck or rail car into a scaling position or area), to be able to sort or upgrade without handling it again would appear desirable. I would like to state clearly, however, that sorting is a step and a cost in itself even if tied to scaling and "merchandising." The highly specialized pressure of sorting to diameter, grade, length, and species is an economic necessity, so much so that some locations use as many as 55 sorts.

More sorting is now being done by specialized machines or by machines that do it better—hydraulic swing cranes and small, fast rubber-tired loaders. Another approach is the mechanical log sorting line. This often combines other values with scaling—such as accounting inventory control and operational sorts for species,

The progressive Totem Sateko log dropout sorting system in use at Evans Products Co.'s Missoula, Montana, operation. Superior Multi-Sort, a Portland, Oregon, company, also manufactures equipment of this type.

Weyerhaeuser Co.'s Longview, Washington, sorting chain. Hydraulic swing crane loads merchandising deck. Barked logs are removed from sorting bays by forklift. Accurate sorting has become an economic necessity.

diameter, length, and grade. It may even include a computerized appraisal of best yield from the log.

Also, some effort is made to obtain a primary or basic sort at the landing or in the woods. Closely connected with sorting is upgrading or piece selection.

Photographs of some sample yards are shown in this chapter. I wish I could show all yards, because each answers an individual need or problem. My intent is not to select one as better than another, but merely to show the concrete steps being taken to meet current problems.

These problems are of two kinds: (1) to handle more logs and to handle them more cheaply using current technology of forklifts and cranes; and (2) to upgrade the basic raw material. It is in solving the second problem—upgrading—that so many factors present themselves: remanufacturing, rebuilding, elimination of bottleneck logs, and so forth. Much is happening in today's mill yards. Are you alert to the opportunities?

In addition to the traditional and functional large forklifts and cranes, some new concepts are being presented for the final mill yard area, the storage area. These include, for example, the Soderhamn-Bush products; the Link-Belt Speeder, Fabco Div., FMC Corp.; and R. G. LeTourneau's jib cranes.

In conclusion, I sincerely hope I have not slighted any manufacturer of equipment or mill yard of distinction. Interesting and

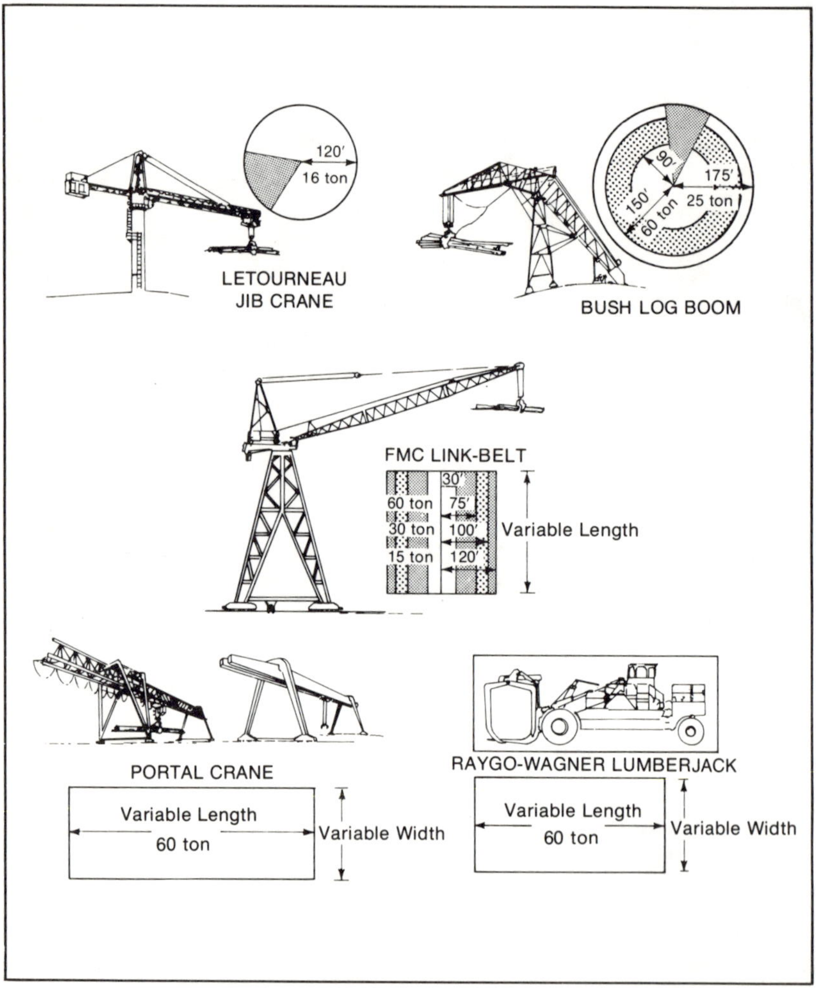

One consideration in comparing the log handling equipment shown is the amount of land required by each for log storage.

productive things are happening in many mill yards and manufacturers' plants. Often two, three, or even five different manufacturers can and will be involved in the presentation of equipment available to upgrade your dry-land mill yard. It is a changing, growing technology. The mill yard is today a true "profit center" that can yield great returns.

DISCUSSION PERIOD

QUESTION: What can be done about tangled logs when moving logs away from a dropout space with front-end loaders or with other equipment?

CORBETT: In any dropout system, the collecting bin or bay commonly causes problems. This is particularly true with front-end loaders, depending on the size and the mouth and the grapples. I have seen many conveyor chains picked up by the hold-downs on these loaders. Solving the problem requires some intelligence and some operational skill. Superior Multi-Sorts, Portland, Oregon, has what I think is a unique pocket concept that anyone considering mechanical sorting should investigate.

2

Upgrading log bucking and sorting methods

Erwin C. Bahlburg, Assistant General Manager-Wood Products, Publishers Paper Co., Portland, Oregon

In 1972 Publishers Paper Co. was considering increasing quad production and debarking capability at its Molalla, Oregon, mill. Publishers investigated various methods and decided upon the Totem-Sateko log sorting system.

The Molalla mill is our best lumber producer, both in terms of cost and production of specific lengths of lumber that command the highest price. Prior to installation of the Totem system, the mill had a small pond and two decks. Production was 200,000 board feet or less per shift from the headrig and the quad. Early in 1974, the Totem system was installed. A computer was added and a chain was run from the end of the Totem system directly into the mill. These changes increased production to a range of 260,000 to 320,000 board feet per shift. The average number of logs going through the Totem, bucked to length, is 2,600 per shift. The maximum has been 3,280. This addition supplies surplus barked logs for the second shift. At Molalla, three additional decks and two additional barkers were provided.

The Totem system is connected directly to the sawmill, with additional infeed decks to the mill. If something is not performing as it should, barked logs can be supplied to the mill from the other decks. The benefit of the Totem system is the additional supply of logs to the mill.

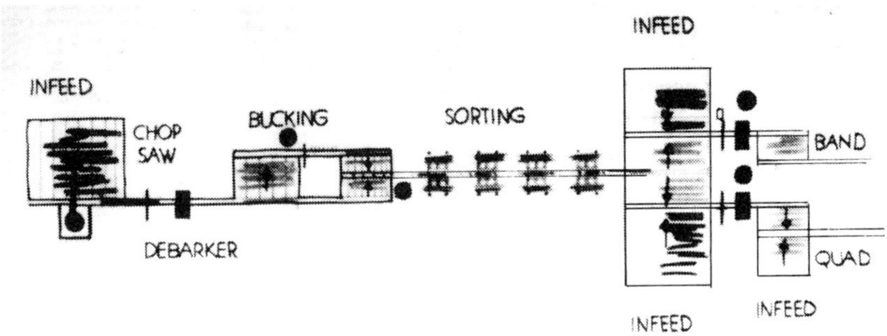

Figure 2.1. Schematic of bucking and sorting operation at Publishers Paper Co.'s Molalla, Oregon, mill.

Figure 2.1 is a diagram of the layout of the Molalla log sorting system.

Sorting system infeed elevated

The infeed deck to the system is 12 to 14 feet above ground. We use a crane to handle logs on the infeed deck and to feed the infeed chain. We have a chop saw following the deck. If we were to do it over again, I think we would use a low-profile drive-on deck and put the bark-carrying refuse conveyors feeding into the main chain in the basement. With the setup we have (being perpendicular to the infeed chain), the operator has difficulty seeing how he is dropping the logs on the chain. We would relocate the crane on the end of the infeed chain. This would improve the operator's view. We would add a kickout conveyor at that point or we would have a scissors conveyor to dispose of crooked logs that would not go through the barker.

We have a 30-inch Soderhamn Cambio barker that will handle logs 4 to 27 inches in diameter.

Computerized bucking increased production at Molalla

The infeed deck to the bucking station will take logs up to 60 feet in length. It is completely computerized. The computer measures the length and diameter in 1-inch increments and then gives bucking instructions to the operator. For example, it might say the log is 54 feet long, it has a 2-inch trim, and it should be bucked for an 18, an 18, and a 16. The operator can override the computer's instructions, but in most cases he does what the computer instructs. The computer is programmed to give the best realization at present market prices, looking at the optimum center yield and bucking for the premium length board that will bring an additional $20 per thousand board feet. This information is kept current by changing cards in the computer.

The computerized bucking operation has increased our average return per board foot considerably. Overrides have been only 30 to 45 per shift. The printout sheet produced at the end of the day records how many logs of certain lengths and certain sizes have

After logs are bucked, they are centered on four 6-inch-high chairs. When the rail collapses it tilts the chair. This deposits the log in the correct bay. The angle of the tilted chair is 22 degrees. Photograph at upper right shows a log coming into position on the carrying chairs.

Totem Sateko system can handle logs from 4 to 30 inches in diameter. This log is about to enter the sorting system. Logs on the infeed deck at left are coming directly from the barker.

been bucked. From this sheet the manager can relate prices of lumber to what has been bucked during the day. Sales return and production have increased. And the length of bucked logs has increased. While no dollar amount can be assigned to this factor, it is evident that it is easier for the mill to handle fewer pieces. The bucking control thereby contributes to increased production. In my opinion, the improvement in the bucking area is most important because it increases the value of the lumber produced.

Speed and accuracy attributes of sorting system

Next we have the Totem Sateko system. The other mechanical operations I have described can be built in any shop. The Totem system takes over after the logs are bucked.

The feeder designed for 4- to 30-inch-diameter logs centers the log on four 6-inch-high chairs. The operator signals from his elevated cab where the log will be released—either to the left-hand side, for the headrig (logs 13 inches and larger), or to the right-hand side, for the quad (logs 12 inches and smaller).

Molalla mill has only eight sorting bays. The Totem Sateko system in use there delivers approximately 10 logs per minute into the bays.

The log sits on the chairs, which carry it down the track. When the log reaches the proper bay, the track collapses, giving the log a quick jolt and thus depositing it in the bay. The logs are not kicked or pushed into the bays. They simply roll off the track into the pockets with the help of gravity. When the release rail collapses, it causes the chairs to slant at a 22-degree angle toward the side to which the log is to be dropped.

Our bays, I feel, could be better constructed, but in most cases they work well. If we had it to do over again, we would construct a decline 6 feet from the rail and then let the logs roll into the holding bays. Once in a while a stacker will hit the conveyor with its fork, causing problems. So I would recommend coming out 6 feet and then a straight drop. Then the machine could not strike the sort track.

The Totem system will take any sweep of log, any crook of log, and deposit it where you want it. We have had no trouble in this respect. The system will take logs from 4 inches to 30 inches in diameter and up to 26 feet in length. It has never failed to deposit a log in the stall into which it was programmed.

The mechanics of the chair system are shown in Figure 2.2. A solenoid is activated. This lifts a trip lever above the top of the

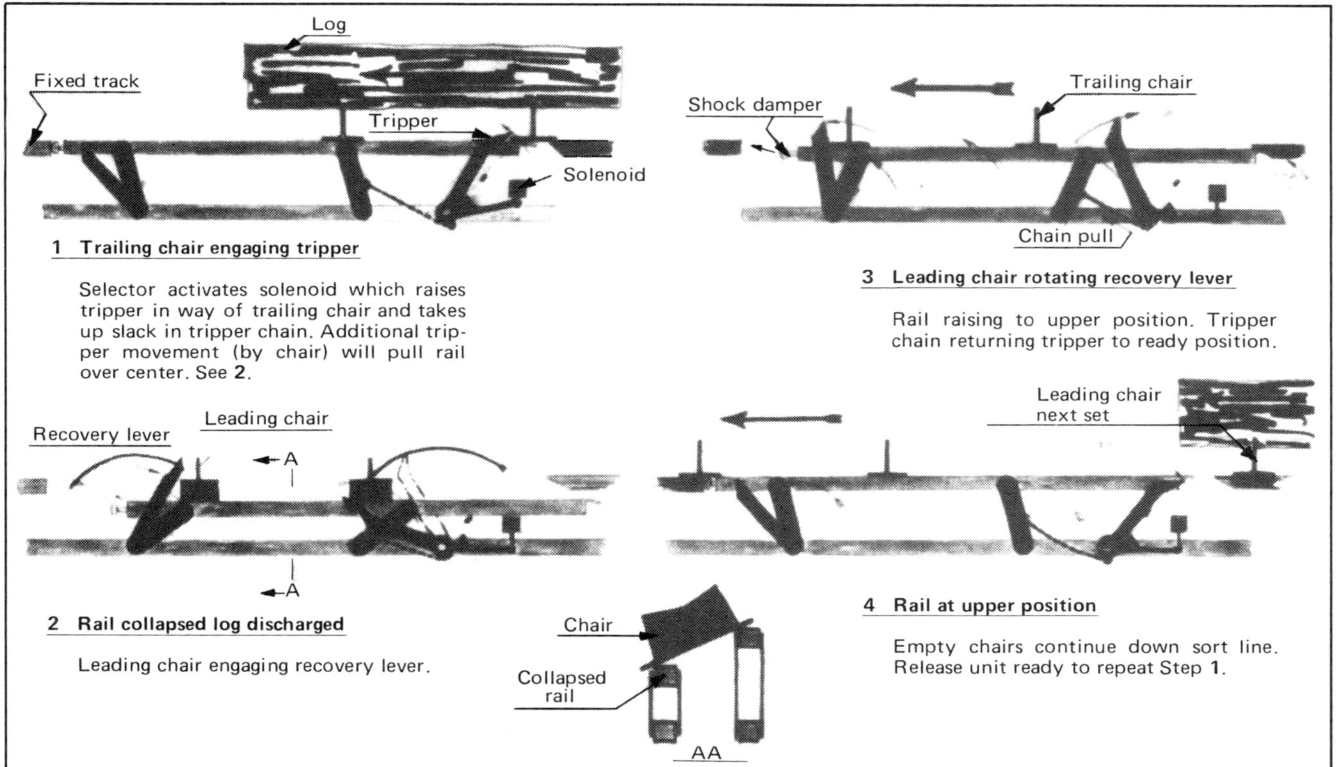

Figure 2.2. Explanation of the Totem Sateko log sorting chair operating sequence.

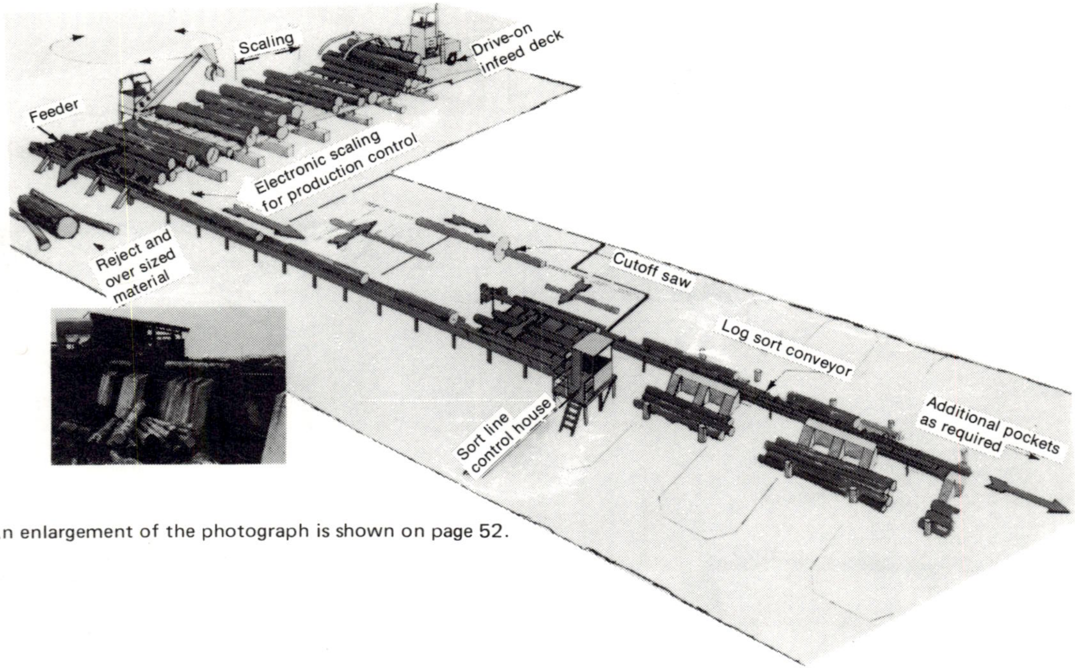

Note: An enlargement of the photograph is shown on page 52.

Figure 2.3. Layout of a typical Totem Sateko log sorting system with drive-on infeed deck.

rail. When the chair supporting the log its the lever, it causes the rail to collapse, thus tilting the chairs and allowing the log to roll off into the bay. The first chair supporting the log, which has been deposited in the bay, resets the rail. So the system works faster than the eye can see. It is completely mechanical, with the exception of the solenoid valve, so it is practically foolproof. The system will deliver 10 logs per minute.

Figure 2.3 shows an example of a low-profile drive-on infeed deck. The crane is on the far side of the infeed deck from the position at Molalla. You can have as many bays as you want—26 sorts, for example. We have only 8 at Molalla, which seems to satisfy our needs for the present.

DISCUSSION PERIOD

QUESTION: Is a special steel used for the runners for the log chairs? I would think they would wear out rather quickly.

BAHLBURG: The runners do wear out quickly, because the chairs move very fast. Totem has built a new release unit. The one we have at Molalla was built for only 4,200 pounds capacity so they have added additional steel to that section. The new chairs are larger and much huskier and the rail is also heavier.

QUESTION: Do you run the rail dry or do you lubricate it?

BAHLBURG: We lubricate the rail.

QUESTION: The first of the mechanical sorting systems I encountered in the 1950s had a kickout for metal detection—a place where the logs containing metal were kicked out. You did not mention anything about metal detection. How do you handle it with your sorting system?

BAHLBURG: We really do not know if the logs have metal in them. If we buy a patch within the boundary of the city, we cut the logs 6 feet above the ground. We have done that in the past when buying logs from close-in areas. If they are not bucked up 6 feet, we will not buy them.

QUESTION: You have a chop saw at the beginning of the system, directly after the hydraulic crane. I did not see it in use in the movie you showed. For what is the chop saw used?

BAHLBURG: Occasionally when there is a log that will not go through the barker easily, the operator cuts it into segments, but he does not use the chop saw very often. I think that saw could be eliminated. The crane could take out those logs that are too crooked to go through the barker, and they could be bucked in the yard.

QUESTION: Who operates the chop saw?

BAHLBURG: The employee on the crane operates the chop saw.

QUESTION: What is the speed of the sorting chain?

BAHLBURG: The outfeed from the Totem is about 280 lineal feet per minute. To put 3,000 logs through a system per day, things have to be moving pretty well.

QUESTION: Who supplied the computer system?

BAHLBURG: Digital Engineering Corp. supplied the computer. It was fairly inexpensive. In fact, I was surprised at the low cost. I actually think a computer will make more money for an operation, for the price, than a number of other improvements, because of its accuracy. And it also provides a printout showing what was accomplished during the day, so you have a record of what the operator is doing and of the number of overrides he is making during the day.

QUESTION: Is that count of 2,600 to 3,200 logs into the bucking system or out of the bucking system?

BAHLBURG: It is the number of logs coming out of the bucking system.

QUESTION: Do you know how many logs you put into the bucking system per day?

BAHLBURG: We get 2 to 3 segments out of each long log, and the number of segments leaving the bucking system is approximately 3,000 per day.

QUESTION: Is there a record of the number of overrides in the system so you can see how it is performing?

BAHLBURG: The computer will print out the number of lengths per day that have been bucked and it will tell you the diameter of each length that has been bucked. This will show whether the logs are running to size and whether the computer is making the right decisions. You have a record of the logs produced; you know which lengths demand an additional $20 for an extra 2 feet. Before installing the computer, the operator was bucking too many segments—instead of three segments out of a log, he might have been bucking four. Since installation of the computer, the log segments have become longer. We are, therefore, putting fewer logs through the sawmill but our production has increased, board footage-wise. Also, with the computer we are theoretically bucking the correct diameter for the size cant which will bring the best price.

QUESTION: What is the length variation and the number of sorts?

BAHLBURG: Log length is from 8 to 26 feet for our eight bins. You can have any number you want, for $18,000 per bin.

QUESTION: What effect has sawing with longer logs had on overrun?

BAHLBURG: I would say it has improved somewhat, but it is difficult to comment on overrun at the Molalla mill, because the mill buys some logs by the pound. I have seen logs 20 feet long, but 2 inches in diameter, thrown in to add weight. We also see some limbs.

QUESTION: You said the computer shows the diameter and the length of the log. Does the computer give a cubic or a Scribner readout at the end of the day?

BAHLBURG: The computer supplies a Scribner readout.

QUESTION: Are you presorting in the yard to balance the flow into the mills?

BAHLBURG: Not yet. It takes quite a bit of room to presort. In Finland every log is 15 feet long, so Totem systems in Finland sort to diameter. Where we have a variety of lengths, it is most difficult.

QUESTION: How do you guarantee you will get enough small logs to run your mill all day if you do not sort in the yard?

BAHLBURG: There is no guarantee.

QUESTION: Are you squaring the logs before they go to the computer-scanner? If not, how do you handle the undercut as far as the readout that comes to the office?

BAHLBURG: We do not square the logs. The readout is quite close even with stump-shod ends or an angular cut. And it will read out whether the butt end or the small end is in front. The operators say they have had little difficulty in that area.

3

Automation: keeping it practical and running

Amos A. Horner, President, Tomco Inc. & All-American Stud Co., Sweet Home, Oregon

I want to share with you some of our experiences and problems in the field of automation. We are living in extremely trying times. Timber supply problems have many of us almost to our knees. We face astronomical stumpage prices, government harassment, a Forest Service plagued with environmentalists and doing an increasingly poor job of managing our resources. In trying to meet the challenge of these times, Tomco has found indeed that automation *can* be practical and it *can* be running. In meeting the challenge, we have found, for example, that we can get a return out of our logs of more than 12 board feet per cubic foot. This helps when you go to the bidding table. The rewards, financial and otherwise, can be tremendous.

This is not the first go-around for us. As have many of you, we failed to make it work the first time. I continue to hear about scanning devices, for example, that have been on the line two years and are still not dependable. Do not be discouraged.

In our case, inoculated with generous portions of "Try Once More" theory–Tomco–we plunged right in to see whether we could really succeed in the arena of maximum yield. The maximum yield approach requires more and more precise machines, possibly controlled by computers, performing tasks such as scanning, charging, setting saws, controlling feeds, sorting, and so forth. Innumerable things are possible. It does not have to be an exercise in futility. Have courage. Plunge on. We did.

If you intend to automate, you must think in terms of priorities. I could certainly tell you many things not to do. But rather, I will talk about the things you should do.

You have a dream, possibly, of building that new sawmill. You walk up and down the aisles of equipment exhibits and look at all the goodies. Couldn't you spend a lot of money if you would just let yourself go? Maybe you just want to automate your edger. Maybe you are going to redo the Scragg mill. But whatever you plan in terms of automation, by all means involve your people early. Number one: involve your people. The performance of whatever you do will depend directly upon the people on your staff. In my case, we were particularly lucky because my son David and his staff are very capable in this field.

After you have involved your people, you should hire professional help. Get commitments from that professional help, the same kind of commitments you have from your own people. Do your research. Look around. Find out what is running. Ask Amos what his system is doing. It might not fit anything you are doing, but you will get an education. Go everywhere. Find out all you can–what kinds of machines do the job, what is available, who seems to be having the most success at the moment. Research, research, research! Put your ideas on paper. Paper is certainly cheaper than components or cutting steel.

Automation: what is it and why do it?

What is automation? A perfect example of automation is the human body, the functions that carry on automatically without thought or action prompting them. A machine that operates in the proper way at the proper time and in an automatic manner is automated, as far as I am concerned. Those functions that must be done on a repeated basis–now, now, now–and those functions that demand extreme accuracy lend themselves particularly to automation.

Why do we automate? We do it to make money. I mentioned the timber supply problem. Believe me, when I go to the bidding table, I have to have muscle to stay there. We all do. The bottom line on our statement is what the bankers look at. Money pays the bills and keeps the store open. Because they are our largest single cost, logs have to be handled in the best possible way. We cannot afford mistakes in cutting. In the old days when we were dealing with large logs and low-cost timber, this was not so critical.

In our stud mill at Cascadia, Oregon, we are dealing primarily with logs less than 12 inches in diameter. The logs vary in size and each must be cut to a slightly different pattern than the one before to get maximum yield. Sawyers obviously cannot decide what pattern is right for each log that precisely and that accurately and do it quickly enough when dealing with 6-, 7-, or 8-second cycles. So your scanning device must work precisely and absolutely accurately, and it has to do it *every time.* It has to perform whether the sun has come up or gone down. It has to perform despite being partially dirty or despite fog or oil mist or dirt in the atmosphere around it. It has to be dependable. It must work every time, without fail.

Design requirements include simplicity and reliability

The way to get production is to keep the log in the saw. If you have the log in the saw and the saw working at maximum efficiency and you are sawing all the time, you have got to be making lumber. You must minimize waste time to get production. To do this, look at each step. Analyze what areas are slowing down your sequence of operation. And then ask yourself, "Can I improve it?" Again, keep the wood in the saw.

First and foremost, an automated system must be reliable. The most accurate system, the fastest system, is totally a waste of time and money if it is not running. If we are running, we are either making money or losing it. If we are not running, we are just plain losing money. A system must have reliability.

From the beginning, when you talk to your people and involve them and as you proceed with your looking, your research, and your planning, use the old kiss principle. Remind yourself, every hour on the hour, "Keep It Simple, Stupid." There is a tremendous tendency for all of us to overcomplicate. Keep it simple.

Other points: Are you designing the system so you can troubleshoot it? Have you built it strong enough? Is it dependable? Are the components you are using off-the-shelf and reliable? I have heard many comments about overbuilding. I have never been able to do it myself. If you do not believe that, build anything as strong as you can, buy a new log stacker, and hire a man to run it. I assure you he will demolish whatever you have built.

Commitments from staff and suppliers vital

I keep talking about people. People are the name of the game. If you cannot involve your people, if they in turn are not challenged by your plans, and if they have not made a commitment to you, you are going to have problems. You will have problems in design. Later, you will have problems in operation. Finally, you will have problems in maintenance.

Another requirement is thorough training of any new people. If you do not have enough people on your staff to meet the needs of the new operation or if you are going into a computerized installation and you need someone with training and background in that field, I advise you to hire that person early in the game. As soon as you find you are committed, hire him. Somehow or other, begin to train him. Then as soon as he is trained, train his replacement, because somebody will hire him out from under you. All these are "do" things. Take my word for it.

You must have knowledgeable, competent suppliers. A supplier who is cheap—or a part that is cheap—is just that. It will haunt you, particularly in the area of automation. Get the very best supplier you can. And if the system requires a computer, I would advise you purchase it from a company with a good, dependable track record with national manufacturing facilities and reputation—a company that can back up the machine with a maintenance program.

Allow for expansion and document the system

When you design this system, remember that we are always building or rebuilding in the sawmill business. When we have nothing else to do, we expand. We see whether we can't run our competitor out of business. This means another shift; it means more machines. Leave yourself room to expand.

The log bucking station at Tomco's Cascadia, Oregon, stud mill, where logs 12 inches and smaller are bucked to 7-, 8-, and 10-foot lengths.

Charger and overhead carriage at Tomco stud mill. A yoke moves the log through the scanner into position forthe carriage to grab it.

When you build the room for your computer, put it on a good foundation. It is best to isolate it from the rest but it is not necessary. The computer room should be large enough for the equipment and the operator and it should be as comfortable as an office should be. It should have visibility—you ought to be able to look from the computer room out to the machines that are being operated by that computer. If the technician can stand in the computer room and see what is going on, it is a tremendous help. Vision is very important.

When you are starting up the machine, remember that safety first is the name of the game. One person should be in charge. The others should look to him for direction. When you have it running, remember to update all drawings and prints. Be sure your engineers leave with you the very latest data. When you make changes, be sure your drawings and data sheets are upgraded and kept current. Document all your systems and methods. Put together written procedures—this is tremendously important in bringing along the next man in line if someone quits or is hired by another company. I cannot repeat too often—document, keep records, know where you are.

Mill flow described

Figure 3.1 shows a flow chart of our stud mill at Cascadia. We start with bucking. We cut 7-, 8-, and 10-foot lengths. We try to get log lengths that are compatible. The logs are measured by a Simco system. Bucked logs go onto a storage deck. The cutoff saw operator also operates the storage deck. Notice that the computer control runs the storage deck, the automatic charging, the scanning, the positioning, and then the sawing and the automatic feed speed. We edge and sort manually, with automatic stacking.

We had to build this mill in the basement of the main sawmill, because of space considerations. We engaged Applied Theory Associates, Inc., at Corvallis, Oregon, and Farwest Equipment & Control Corp., at Eugene, to help us plan this installation. Bill Bolton, Larry Hunter, and the staff at Applied Theory and the staff at Farwest collaborated with us in rebuilding the operation at Cascadia so it would work. We use a Hewlett-Packard computer; we have Opcon scanners; the machinery is by Farwest and Tomco.

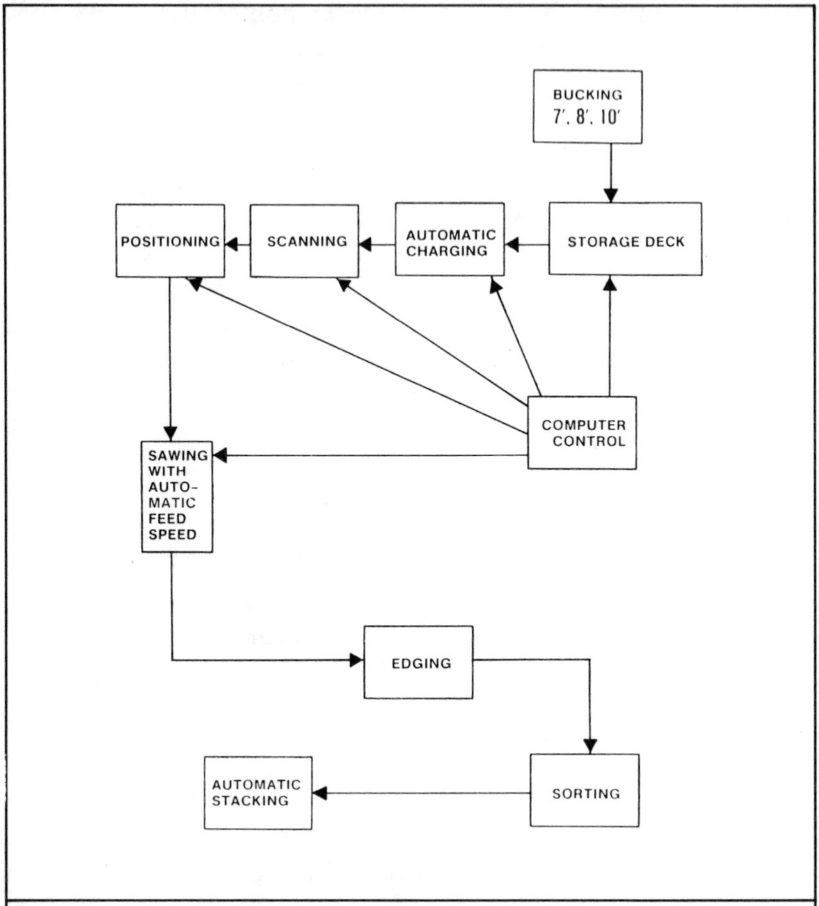

Figure 3.1. Flow chart of Tomco Inc. stud mill at Cascadia, Oregon.

At Cascadia, we have a conventional Scragg mill with an overhead carriage. The mill has two stationary saws, two movable saws, and two chipperheads. We move the chipperheads very precisely. A yoke moves the log in through the scanner and the overhead carriage grabs the log when it is in position.

The computer room includes the Hewlett-Packard computer; a typewriter called a Terminette; and a cathode-ray tube, similar to a television tube, that gives us visual information. I call this the question box. We can ask it questions and it will recall various types of information on the status of the system.

Computer room at Cascadia includes a Hewlett-Packard computer, a Terminette typewriter, and a cathode-ray tube for visual display of information collected by the system or requested from it by the operator.

Studs leave the mill on rollcases and go to the automatic stacker.

Data from system important management tool

Figure 3.2 shows one of the reports we can request from the computer. It is an ongoing readout that tells diameter, length, and time. The right-hand column shows how many seconds it took for each log. We monitor these things very carefully. In fact, the computer operator keeps a record of our downtime. If downtime exceeds 25 seconds for any reason, he records it. It might be a problem at the edger, but it is extremely important that management know these things. We constantly use these daily records as a tool to increase the time the machine is available on the line.

We also can ask the computer what the previous log's calculations were, to check its performance. Let us start at the top of Figure 3.3. The report shows the diameter, length, and speed code (I think we have eight different speeds); the taper set; the required skew and the way it was set; where the chippers were set; where the saws were set; and finally, the yield from the previous log. This particular log was bigger (13 inches) than we like because recovery is not as good from the larger logs. That is one reason we stay below 12 inches primarily.

Figure 3.4 shows an example of an accumulative report. The machine can be asked to type out this type of report any time you want to know what you have done. The most significant thing shown in this report is that the projected lumber tally was 81,485 board feet and the actual tally produced was 81,490—that is coming very close to the estimate. The piece count shown on the report also is very important to us, because we are not computerized at the edger. We do not have an exact means of determining whether we have edged properly, as it was calculated to be edged. But by using the piece count on this report, we have a check as to our efficiency. This is very important. The accumulative report is available for each shift and is an important management tool. You can check through the planer and elsewhere down the line to be sure the numbers agree.

11/21/75	14:15:23	DIAMETER	6.1	LENGTH	8	9	SECONDS
11/21/75	14:15:31	DIAMETER	6.2	LENGTH	8	8	SECONDS
11/21/75	14:15:39	DIAMETER	6.2	LENGTH	8	8	SECONDS
11/21/75	14:15:48	DIAMETER	6.0	LENGTH	8	9	SECONDS
11/21/75	14:15:56	DIAMETER	5.0	LENGTH	10	8	SECONDS
11/21/75	14:16:04	DIAMETER	7.9	LENGTH	8	8	SECONDS
11/21/75	14:16:13	DIAMETER	7.5	LENGTH	8	9	SECONDS
11/21/75	14:16:21	DIAMETER	6.1	LENGTH	8	8	SECONDS
11/21/75	14:16:29	DIAMETER	5.4	LENGTH	7	8	SECONDS
11/21/75	14:16:38	DIAMETER	9.1	LENGTH	10	9	SECONDS
11/21/75	14:16:46	DIAMETER	8.4	LENGTH	10	8	SECONDS
11/21/75	14:16:54	DIAMETER	4.5	LENGTH	8	8	SECONDS
11/21/75	14:17:03	DIAMETER	10.4	LENGTH	10	9	SECONDS
11/21/75	14:17:11	DIAMETER	8.6	LENGTH	8	8	SECONDS
11/21/75	14:17:19	DIAMETER	7.4	LENGTH	8	8	SECONDS
11/21/75	14:17:27	DIAMETER	6.7	LENGTH	8	8	SECONDS

Figure 3.2. Computer readout monitored to control downtime.

11/21/75 15:24:32

DIAMETER S	13.0	IN
LENGTH	8	FT
SPEED CODE	1	
H TAPER	0	IN
R SKEW	0.539	IN
S SKEW	0.539	IN
C OFFSET	0.379	IN
S OFFSET	0.379	IN
CHIPPERS		
LEFT	5.505	IN
RIGHT	5.505	IN
SAWS		
L FRONT	3.665	IN
L REAR	1.825	IN
R REAR	1.825	IN
R FRONT	3.665	IN

LAST LOG CALCULATIONS

LUMBER TALLY	6.02	CUB FT	90.7	BRD FT
LOG VOLUME	7.29	CUB FT	50.0	BRD FT
RECOVERY	82.57%		181.40%	

PIECE COUNT

	NUMBER	BF
2 x 4	17	91

LAST LOG YIELD

Figure 3.3. Report used to check performance of computer.

TOMCO INC., CASCADIA, OREGON

11/23/75 17:27:50

SHIFT DAY FOREMAN CLYDE

PROJECTED LUMBER TALLY	5127	CUBIC FEET	81485	BOARD FEET
LOG SCALE	6642	CUBIC FEET	33410	BOARD FEET
RECOVERY	77.19%		243.90%	

BF/CF = 12.26

NUMBER OF LOGS RESCANNED	46
NUMBER OF REJECTED LOGS	3
NUMBER OF LOGS SAWED	1878

LUMBER PRODUCED PIECE COUNT

	6 FT	7 FT	8 FT	10 FT	BOARD FEET
1 x 4	79	144	1050	132	3734
2 x 3	53	41	348	41	1900
2 x 4	56	1311	10784	1800	75856

Figure 3.4. The shift report—an important management tool.

System can pinpoint malfunctions

This equipment has the ability to do a tremendous job of self-analysis. This is important because time is money. The machine will keep us informed, if we ask it, about what is going on. If a hose or a limit switch were to break, the self-analysis feature would point out where to look for the problem.

When operating equipment of this type, it is necessary to develop a new level of awareness. In maintenance, for instance, we should be far cleaner and more organized than we have been in the past. Maintenance people should be scheduled to do regular things at regular intervals. Precise machines that have to be properly adjusted must be checked regularly and should be under continuing surveillance. We must get rid of some of our old habits. Yesterday's level of performance is no longer adequate. This is not to say that in the new mode we will work harder. We will only work differently. The job today is far more challenging and far more involving—and it is tremendously important and significant to the well-being of your company and your people.

One last thought: Do not automate with the idea that you will save people. In this particular installation we have exactly the same number—perhaps even one or two more—we would have if we were not automated. Automation does not necessarily reduce the number of people needed, it upgrades the quality of the work you do.

Finally, I would be remiss in not mentioning that Applied Theory Associates and our construction people have been particularly good. When we hired them (and this is important; it is something you should try to do) we got commitments from them. For example, my commitments with Bill Bolton of Applied Theory were that we did not finish payment on the equipment and the automation devices until the system had run one full week without a minute of downtime. Further, it had to run 30 days with 95 percent availability. Further, it had to stay on the line 6 months with complete guarantees. I would admonish all of you—make your deals on some similar basis. Do not get trapped by fly-by-nights.

What do I think about automation? I am tremendously impressed. It has been a beautiful experience. I sometimes cannot believe the numbers, but they are accurate. We have checked and double-checked them.

I might add that our charger, which has a ballscrew setworks, sets very precisely all day and at a very minimal thousandths variation. A charger and setworks very similar to the one at Cascadia has been built by Farwest for Simonson Lumber Co., at Smith River, California. The third Farwest builds will go to All-American Studs, Springfield, Oregon. We are also putting a scanner and computerized system on the new Letson & Burpee double-cut headrig that is going into All-American. We plan computerized scanning of all the logs going into our new veneer plant at Sweet Home, Oregon. It will go on line as one of the first, if not *the* first, computerized scanning-charging systems in the veneer business. We also plan a long log mill at Sweet Home. We intend to take advantage of all the technology available to make that mill the very latest and the very best.

Comments by Bill Bolton, Executive Vice President, Applied Theory Associates, Inc., Corvallis, Oregon

I want to express my appreciation to Tomco Inc. for the opportunity to do the mill at Cascadia. I would like to add a few thoughts to those Amos offered. It is terribly important as a supplier to this industry that we have the opportunity to involve the mill operating people early in the program. I cannot stress that enough. We are applying technology which is different—in some cases, radically different—from that to which the operators are accustomed. Your people have to operate the program and the system on a long-range basis. Obviously, the supplier is on call, but he is not there operating the equipment on a 16- or 24-hour basis. It is very important that the personnel from the lumber company ask plenty of questions, no matter how trivial they may think those questions are. This is not a situation in which to be embarrassed or hesitant about asking any question.

DISCUSSION PERIOD

QUESTION: Do you find you must have a full-time data processing manager or someone in the computer room at all times?

BOLTON: It is not necessary to have someone running the computer at all times. It *is* necessary that the personnel be trained to initiate the program and to understand the data coming out. They must be knowledgeable as to what they can call up through the keyboard terminals. The cathode-ray tube, or CRT, puts out on its screen information that monitors all the functions in the mill—not just the setworks, but all the limit switches and the scanners—on a continuous basis. If there is a hangup in the mill, that program goes into an immediate diagnostic and points to the problem area. It is important that mill personnel and operators be able to understand the system.

QUESTION: Do you use the computer in the system for general data processing—accounts receivable, accounts payable, or general inventory control?

HORNER: At this time we are not using the computer for other functions. We plan, however, to integrate it into such things as inventory and business functions, using it on an off-time basis.

I doubt we will try to do it while the mill is running.

QUESTION: You plan to use the same computer, then?

HORNER: That is affirmative.

QUESTION: How did you get into this field initially? What kinds of questions did you ask yourself, to be the leader? You implied that, except for your son, you had nobody in house who had background in this area.

HORNER: Seeking the ultimately competitive sawmill, I do not think there is a man in the sawmill business who is not always thinking about how he can improve or rebuild. Probably because of my son David and some of his staff having talent in this direction, it was easier for me to venture into it. Obviously, attending Sawmill Clinics and hearing about and seeing operations with scanning devices and computer arrangements makes it almost impossible not to be challenged. That is what sawmilling is all about—it is a great challenge.

QUESTION: Did you say you get 12 board feet per cubic foot?

HORNER: Actually, we get more than 12 board feet per cubic foot.

QUESTION: How do you measure log diameter? Do you take the diameter at the small end and then take a right cylinder and calculate it?

BOLTON: It is based on the diameter at the small end, but we do not use a right cylinder. We measure the distance across the log at frequent intervals—at Cascadia, it is at 4 9/16-inch intervals. This gives us the diameter across the log through its full length. We then compute the taper. What we are sawing is a tapered cone.

COMMENT: So you do calculate on a tapered cone and not on the small-end right cylinder?

BOLTON: That is correct.

4

Modifying a mill to fit today's logs

Tom Fleming, Superintendent of Lumber Operations,
American Forest Products Corp., a subsidiary of
the Bendix Corp., Martell, California

The problem of what to do with a large-log mill, as the changing log mix begins to affect productive capacity, has been before us for some time. I address this audience with considerable trepidation, knowing that it includes most of the best sawmill people in the world. On the other hand, the way we are approaching this may be of interest.

Much thought is being given to the influx of more and more small logs into the log mix throughout the sawmill industry. What to do with this ever-increasing input, while maintaining a reasonable rate of return, has become of marked interest to virtually every lumber manufacturer. Most have taken steps of one kind or another to cope with this difficult situation. Note the number of operations participating in the export market and those with combination systems or new small-log plants operating or coming on stream. Weyerhaeuser Co., Boise Cascade Corp., Willamina Lumber Co., U.S. Plywood Corp., Paul Bunyan Lumber Co., and Kimberly-Clark Corp., to name a very few, all employ different means in attacking and solving the problem.

To maximize the benefits of our forests today, management must be intensified to obtain optimum growth. This requires, among other things, more commercial thinning, as well as utilizing to a smaller top diameter. The end result is a far greater number of logs developing in the 4- to 12-inch class than heretofore. While we at Martell have not done much in this respect so far, we have become concerned with the increasing numbers of smaller logs showing up at the headrigs at considerable cost to our productive capability and overall return. Smaller logs mean lower lumber recovery factors and less return. So we must become even more efficient than we now are.

Fire emphasizes need for studies

While we had been aware of this for some time, in August 1973 a destructive, costly forest fire, starting along Highway 50, raced through some federal timberlands and into a large stand of our young-growth timber, killing nearly 60 million feet overall. We ultimately logged all that timber. This created an instant supply of small logs. It hurt. Figure 4.1 illustrates how our log size has decreased since 1965, and what we predict to 1985. In 1966, our average log contained 424 board feet; in 1973, it was 318 board

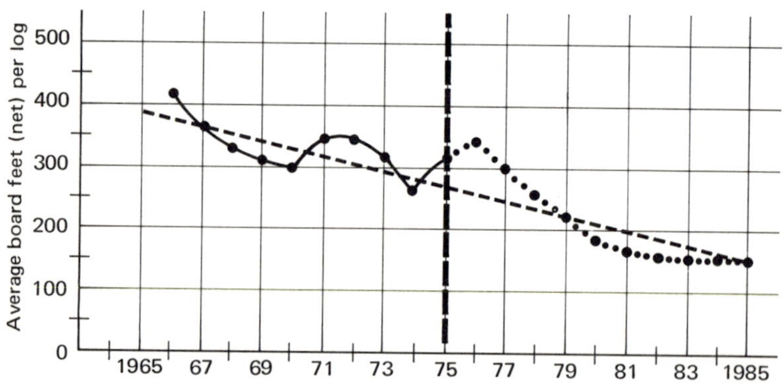

Figure 4.1. Average log scale (1965 through 1985) at American Forest Products Corp.'s Martell, California, large-log mill.

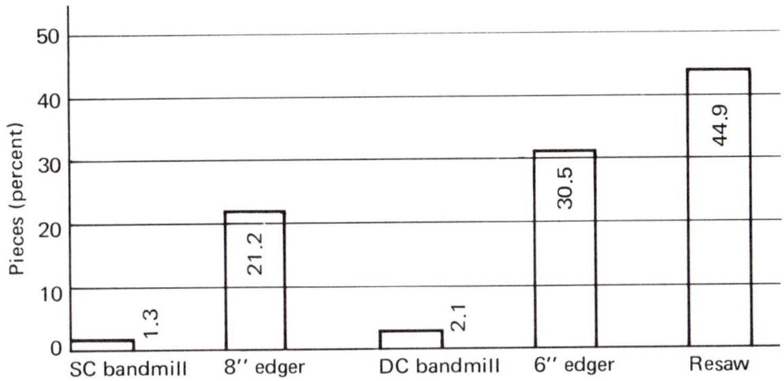

Figure 4.2. Source of lumber pieces presented to the trimmer.

feet; and in 1974, because of the fire, it dropped to 253 board feet. The average has risen since, but the trend is downhill. The real question is: How are we to contend with this problem of more small logs?

At that point, there was little we could do but forge ahead and hope for the best. This did dramatically emphasize, however, that we lacked a great deal of information as to what was needed or, for that matter, just what was happening throughout the mill. We had already begun an exhaustive survey of our timberlands. Our forestry department is breaking this into diameter classes to enable us to program the mill flow better. We have also conducted countless other studies to determine loads at various machine centers. Figure 4.2 illustrates one of those studies.

A few years ago, we made a machine center piece count, which we are currently updating. Then, 1.3 percent of the pieces hitting the camelback ahead of our trimmers was from the single-cut headrig. The big edger from that same side discharged 21.2 percent. The double-cut bandmill inflicted 2.1 percent of the pieces upon the camelback; the 6-inch edger, 30.5 percent; and the resaw, 44.9 percent. You can see where we were concentrating—and where we still concentrate today, I might add, although not as much as some mills.

Our present log mix is 70 percent pine; the balance is white and Douglas fir, although mostly white fir. Figure 4.3 shows the flow in our present plant.

Bandmill flow detailed

We have one infeed deck for plywood peeler logs (1) and one for sawlogs (2) so the operator can mix the logs as necessary to keep both operations going. We supply our plywood plant with peeler blocks and our sawmill with sawlogs from the same barker and bucking saw. We start with a 60-inch Salem-Brunett ring barker (3). We run the barker three shifts, so we can keep the plywood plant going. That plant peels as many as 900 blocks per shift and the sawmill has cut up to 700 per shift. We have an 84-inch horizontally operated circle bucking saw (5), with a 7-foot L-M chain saw horizontally opposite it (6). Logs are then transferred to the sawmill (7) or peeler blocks sent on to the plywood plant in a straight line (8).

The 84-inch cutoff saw is one of our bottlenecks. The action is a little slow. When we are cutting peelers, we cannot get them out fast enough. We have no problem at the barker, but we are constantly delayed by the bucking operation. This is one cog we are at work trying to smooth out.

Figure 4.3 shows that logs for the sawmill travel quite a distance before they get to the headrig. I timed this trip; it took 6 minutes for a log to get from the barker into the headrig. We had rather slow conveyors in the past. Our engineer, Cliff Jackson, is making great strides in changing this line to bring us up to our designed capability.

We have a 9-foot single-cut Filer & Stowell bandmill with a 72-inch carriage (10) and a 9-foot double-cut Filer & Stowell bandmill with a 54-inch carriage (17). Then we have an 8-inch Portland Iron Works edger (11) and a 6-inch Klamath Iron Works edger (12). At the single-cut headrig, we have a so-called Hoover handshaker, which helps hold the wide lumber. Not having a tail sawyer, we use a partial slat bed and rolls. Cliff Jackson designed spin rolls that help position the board so it reaches the edger in the proper configuration.

At both headrigs, we have Beloit-Passavant slabber-chippers—a 36-inch at the single-cut headrig and a 24-inch at the double-cut.

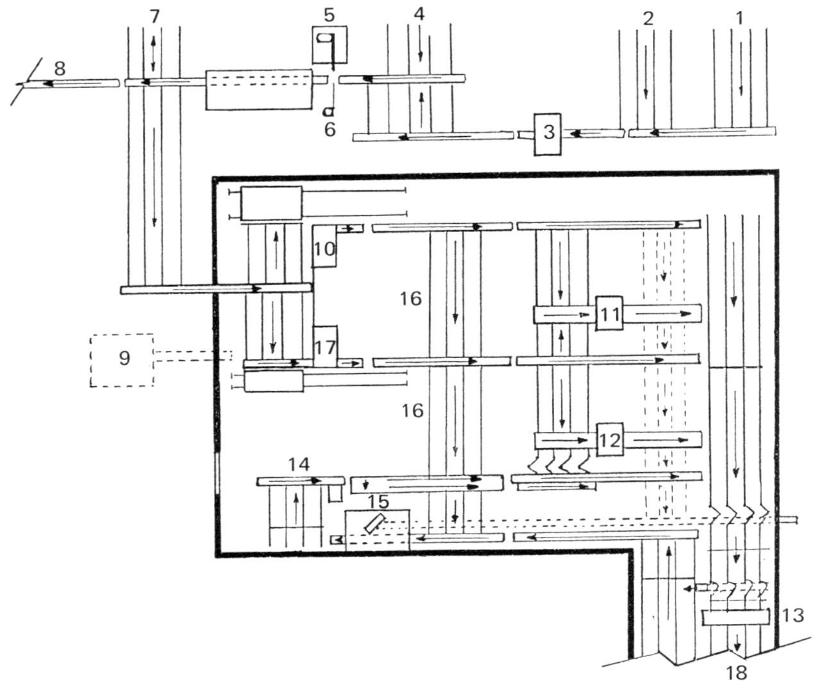

1. Infeed deck (peelers)
2. Infeed deck (sawlogs)
3. Ring barker
4. Bypass deck
5. Circle bucksaw
6. Chain bucksaw
7. Infeed to headrig
8. To plywood plant
9. Transformer bank
10. Single-cut bandmill
11. 8-inch edger
12. 6-inch edger
13. Trimmer
14. Resaw
15. 60-inch CM&E chipper
16. Cant transfer to resaw
17. Double-cut bandmill
18. To Sorter

Figure 4.3. Present flow at American Forest Products Corp.'s Martell, California, large-log mill.

The resaw, a 7-foot Klamath with linebar and C&D setworks (14), was taken from our old sawmill. We have an Irvington-Moore 20-foot trimmer (13) and a Moore Oregon 24-tray sorter. I would rather not mention the stacker.

Since hindsight, it seems, is always better than foresight, we have found it necessary to make many changes in the mill to smooth out the flow. Among these are improved log-conveying speeds; transfer of unedged boards from the resaw to the 6-inch edger and transfer from the double-cut bandmill to the 8-inch edger; rebuilding the resaw feed; adding the two Beloit slabbers; and other changes. These alterations have not only improved the overall flow, but have increased our piece count significantly. Up to now, we have been primarily concerned with improvements in production, sawing for grade as highest priority. In chapter 11, Bob Mardock of Harwood Products explains how they took their mill, which is an old-style mill more or less, and made it saw accurately, efficiently, and with greater production—and still get grade. We, too, are concerned, as log diameters continue to decrease, both with productivity and accuracy. What then were we to do to overcome these problems?

Sizing up future raw material

Ed Williston, of Weyerhaeuser Co., recently stated that there are small logs, and then there are "California" small logs. Those of you who are daily faced with top diameters of 2½ inches and 20,000 or more pieces per shift can move on to the next chapter, for we are, in fact, dealing with the "California" small log.

At the present time, we cut 63 to 65 million board feet in the sawmill annually. (We could cut more, and we are gradually increasing this cut, but we are limited on the basis of sustained-yield parameters.) This production means short log counts of from 400 to 700 per shift daily, with about 8,000 pieces of lumber as an average. Carl Mason, of H. C. Mason & Associates, once said that to obtain a reasonable rate of return at a small-log mill, 40 to 50 million feet of logs would be needed each year, at least. To justify a separate small-log facility at Martell would require all of that. Coupled with the needs of the large-log mill and the plywood plant, this would mean an annual cut of 140 million feet, which is just not available on a sustained yield basis from our timber supplies. We do believe we can plan on an additional 6 to 10 million feet annually from fee and federal lands for some years through commercial thinnings and by reducing top diameters to 4 inches. These constraints forced us to take a hard look at what alternatives were available.

I will admit that I am not adverse to stealing ideas. Much time has been spent in the past few years determining just what has been done throughout the industry—through mill tours, sawmill clinics, technical sessions, Forest Products Research Society meetings, and so forth. Our thoughts have run the gamut from doing nothing to building a separate facility. As far as capital outlay is concerned, the least expensive solution is to do nothing, a solution which would lead eventually to going out of business. At the other extreme, and the most costly alternative, is the separate side. Beside being costly, however, the second alternative is in our opinion redundant, considering our existing and future log mix.

Considering possibilities for modification

With this in mind, we narrowed the possibilities to four which we felt warranted consideration. They are summarized in Figure 4.4. First, thought was given to increasing log conveyor and

1. Increase log conveyor speeds and decrease bucking times. Replace 8-in. edger with an 8-in. by 60-in. combination edger.
2. Add a separate barking and bucking side, and replace 8-in. edger with an 8-in. by 60-in. edger.
3. In addition to item 2, move the double-cut to the single-cut side and replace the single-cut with a canter twin or quad in the previous double-cut position.
4. In addition to item 2, add a separate canter twin or quad to the existing flow by introducing it into the mill between the double-cut and the resaw.

Figure 4.4. Alternatives under consideration for modifying the Martell large-log mill to handle increasing volume of small logs.

bucking speeds in conjunction with replacing the present 8-inch edger with an 8-inch by 60-inch combination battery-selective type. Our second possibility was to add a high-speed barking and cutoff line to the mill, which would also supply most of the peelers to the plywood plant. In addition to this line, we would also install the new 8-inch combination edger. Third, in addition to replacing the present edger as in number 2, we would move the double-cut to the single-cut side and put in a canter quad to replace the single-cut bandmill. The fourth possibility leaves the present bandmills where they are, but replaces the 6-inch edger with an 8-inch, battery-selective type and introduces a new canter twin or quad line between the existing double-cut and resaw bandmills. All methods proposed have their respective advantages and disadvantages and we are faced with many problems, not the least of which is the downstream effect of an increased piece count.

You can see in Figure 4.3, the original mill flow chart, that we have some real problems implementing any of the alternatives. We have 18 feet between posts between the double-cut headrig and the resaw. In introducing our other factors, we can come in from outside, across the transformer bank and then into the mill. We would have room, but we will have problems taking care of the discharge. I am confident though that with Cliff Jackson's experience to draw on, we will be able to resolve them.

Figure 4.5 shows the third alternative listed in Figure 4.4. In this plan, we have moved the double-cut (17) to the single-cut side and put in the canter quad (20). We are transferring the boards from the quad to the big edger (11) and to a chipping edger (21) and then sending the cants over to the little edger (12), which in this case would be a combination type. Since we prepared this plan, however, we have decided to eliminate the chipping edger. We do not think we need it, but we might at some future date.

Figure 4.6 shows the fourth alternative. It includes addition of a new log barking and bucking system (19). We are planning to add two chop saws as well as a high-speed small ring barker, mainly to take care of the plywood plant and at the same time supply the sawmill with small logs. You will notice we have a long transfer. We have a real problem because there is a 12,000 kilovolt-ampere transformer bank (9) under that transfer. I shudder to think what

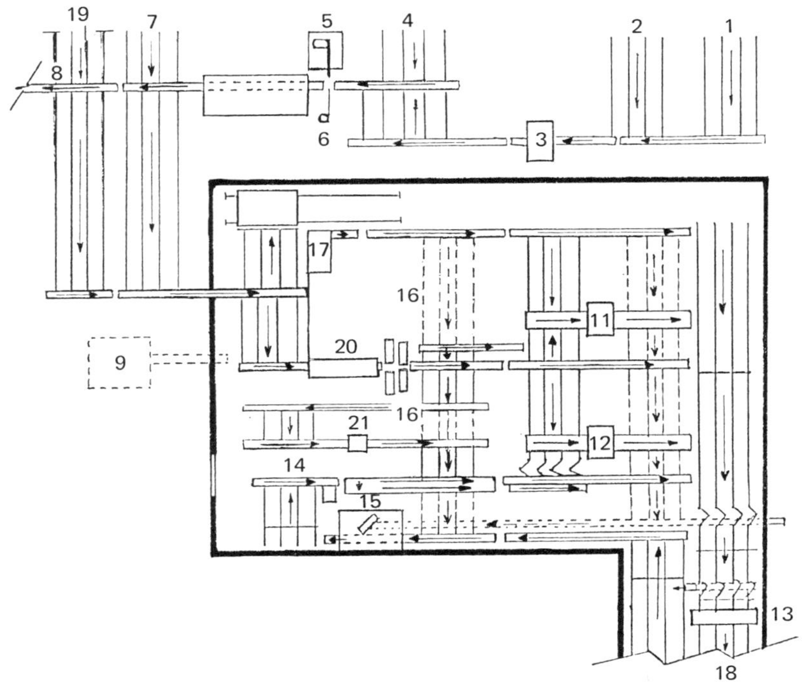

1. Infeed deck (peelers)
2. Infeed deck (sawlogs)
3. Ring barker
4. Bypass deck
5. Circle bucksaw
6. Chain bucksaw
7. Infeed to headrig
8. To plywood plant
9. Transformer bank
10. Single-cut bandmill
11. 8-inch edger
12. 6-inch edger
13. Trimmer
14. Resaw
15. 60-inch CM&E chipper
16. Cant transfer to resaw
17. Double-cut bandmill
18. To Sorter

19. Separate barking and bucking line
20. Canter quad
21. Chipping edger

Figure 4.5. Proposal for mill modification includes adding a separate barking and bucking line (19), moving the double-cut (17) to the single-cut side and adding a canter quad (20) to replace the single-cut. A chipping edger (21) would be used in conjunction with the canter quad.

1. Infeed deck (peelers)
2. Infeed deck (sawlogs)
3. Ring barker
4. Bypass deck
5. Circle bucksaw
6. Chain bucksaw
7. Infeed to headrig
8. To plywood plant
9. Transformer bank
10. Single-cut bandmill
11. 8-inch edger
12. 6-inch edger
13. Trimmer
14. Resaw
15. 60-inch CM&E chipper
16. Cant transfer to resaw
17. Double-cut bandmill
18. To Sorter

19. Separate barking and bucking line
20. Canter quad
21. Chipping edger
22. Possible add-on for resaw

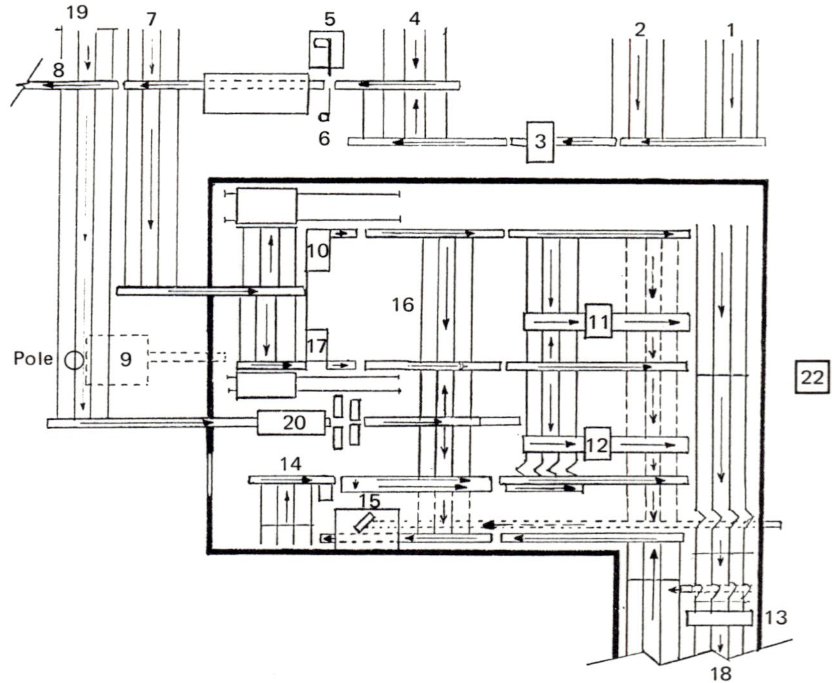

Figure 4.6. Proposal for mill modification includes adding a separate barking and bucking line with infeed directly to a canter quad (20) added to mill between present double-cut headrig (17) and resaw (14).

would happen if a log were to fall on it. The power pole, which is directly behind the transformers, will have to be moved and something will have to be done to protect the transformers. In this illustration, we have brought in the canter quad (20). Since preparing this drawing, I have learned it would be better to move the canter quad outside, just past the transformers—and there is ample room—rather than place it as shown. Then we could transfer to the resaw or disburse the lumber elsewhere.

We investigated one other possibility, also shown in Figure 4.6. It was to take out the resaw (14) and put the quad (20) in its place. We would then have put the resaw at the back end of the plant, outside the present building at (22), and establish a flow back in to the camelback to the sorter.

We know we have the capability of handling more than 21,000 pieces, and we feel certain we can handle 17,000 pieces reasonably well, but we do not anticipate reaching that level within the next three or four years. We figure it will be about ten years before we reach that piece count.

Changes will be made step-by-step

We have received the wholehearted cooperation of the machinery manufacturers and sales representatives we have approached, and many sound ideas have been advanced. At the present time, we favor the fourth approach (Figure 4.6) to handle from 1,000 to 1,500 logs per shift, with piece counts to 14,000 or more. We do not intend to implement this all at one time, however. Our first step will be to replace the edger; then we will put in the high-speed barking-bucking line; and, finally, the canter twin or quad will be added.

Doing things by this piecemeal approach involves little production downtime. It also affords us the opportunity to study and resolve the problems that may arise on a step-by-step basis, enabling us to keep up with the increasing numbers of small logs as they develop. Payouts, too, can be more effectively finalized as improvements in productivity and accuracy materialize. With such a system, we can continue to process efficiently logs from a minimum diameter of 4 to a maximum diameter of 72 inches, from 8 to 20 feet long, in order to maximize lumber yield in the

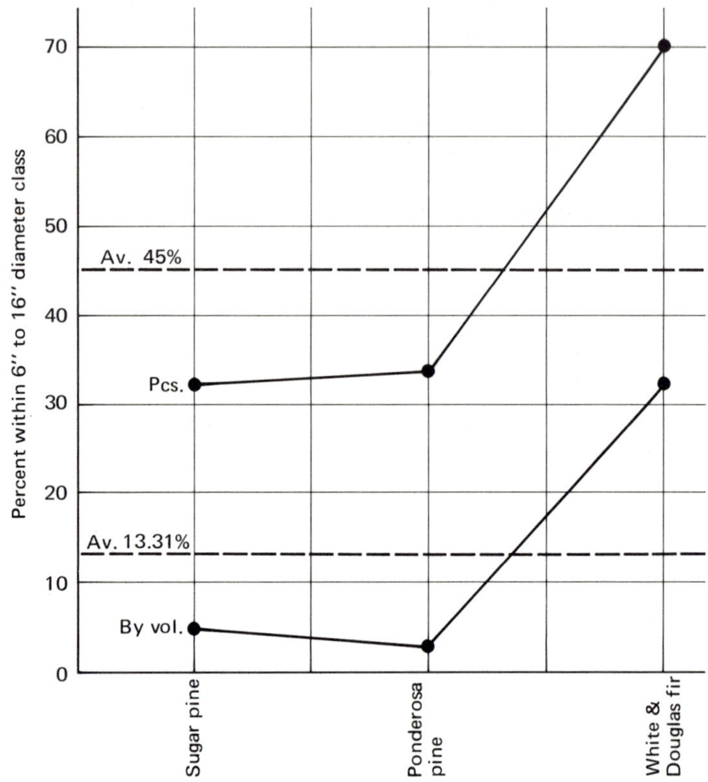

Figure 4.7. Percentage of logs now being sawn (by species) within the 6- to 16-inch-diameter class at Martell.

best dollar return. This flexibility is a must under such conditions.

We are still faced with many questions, such as whether we will be able to handle the number of log sorts required, and whether we can justify the investment needed with the necessary payout. But one final thought: Figure 4.7 shows the situation facing us now. In sugar pine, we are cutting 32 percent of the logs from 6 to 16 inches in diameter, while the volume is only 5.73 percent. Ponderosa pine equals 33 percent, for 3.99 percent of the volume; and the white and Douglas fir combined total 70 percent of the pieces, but 30.3 percent of the volume. Thus an average of 13.31

percent of our volume is in logs 6 to 16 inches in diameter, but 45 percent of the pieces we handle comes from those logs. Based on this information, it looks as if the day of reckoning has already arrived.

The alternatives enumerated are but a few of those examined, and we are fully aware that they may not necessarily be the best available. We do believe, however, that for our mill and log mix, they come closest to fitting our requirements, both with respect to space available and footage to be processed. We are now obtaining quotations on the 8-inch by 60-inch single-arbor, combination battery and selecting edger to be submitted for budget approval. We hope to see its installation in fiscal year 1977, which begins October 1, 1976.

5

A way to keep your small-log side simple

Ray C. Swanson, President, Swanson Bros. Lumber Co. Inc., Noti, Oregon

It is said there is nothing new under the sun. I suppose this truism, like others, is mostly correct. Many of us continue to search for different and better ways of doing things and we sometimes succeed. But nearly everything we do is an adaptation, in some degree, of a process that has been used before or that has existed from time immemorial. The most difficult thing to do, especially in this age of computerized automation, is to develop a process that is simple, efficient, and economically practical. Indeed the computer, that classic epitome of simplicity, where the push of a button will set in motion an infinite variety of operations, may well become a mountain of complexity because chaos occurs if the button fails to work or, worse yet, sets off the wrong chain reaction.

I am not saying that computerized automation is the wrong way to go. I am only prefacing my remarks with these statements to explain why the operation I will describe may not compare with more sophisticated installations. It is also another way of saying that it obviously does not make sense to install a system that will eliminate six lumber handlers from the sorting chain if two electricians and four millwrights are needed to maintain that system. That is the premise from which we begin. I will illustrate the processes which in our opinion contribute to a simple, reasonably efficient, and economically practical installation within the natural limitations of our particular situation.

Raw material variations add complexity

The operation begins with the log sorting yard. In this day and age of extreme competition for raw material, simplicity must sometimes yield to economic pressures that add to the complexity of an operation. Log loads vary extensively in size, length, species, and grade. Hemlock, white fir, old growth, second growth, chinquapin and cedar may all come in on the same load. This intensifies the need for space, speed, and versatility. Also required is a more general understanding of species characteristics for proper identification. It makes necessary a knowledge of markets and utilization alternatives by the entire timber supply division. It is obvious to the most inexperienced and uninformed observer that a uniform log (in terms of size, sweep, taper, and length) would make possible the ultimate in simplicity in the lumber manufacturing end of the operation, if it was feasible to achieve that kind

A Caterpillar 966 is used to sort and transport logs in Swanson Superior Forest Products' mill yard at Noti, Oregon. Diversity of log sizes, lengths, and species requires operator knowledge of economics and utilization.

of sorting performance. But to the logger, the log buyer, the machine operator in the log yard, and the manager, it is equally obvious that the tradeoff in space requirements and machine and manpower investments, as well as the logistics of handling and storage, limit the practicality of this approach. Admittedly, a greater degree of log sorting than we practice would certainly be handier and perhaps more efficient but the decision to work with one machine, a Caterpillar 966, was dictated by space limitations as well as manpower-investment-return considerations.

Since the existing cant-producing facilities are limited to an approximate 15-inch maximum diameter for the round log entry, most sorting is done prior to the pickup point at the No. 1 Ramey grapple loader which feeds the infeed slip. Ahead of the debarker, there is also an oversize log kickout controlled by the chopsaw operator, who makes a visual judgment of diameter and sweep limitations. The log then enters the debarker, a 26-inch-capacity Valon-Kone ring machine. At this time, we are processing nothing bigger than a 15-inch-diameter log. For logs with small deviations in excess of the 15-inch diameter, a reducing ring device in the

Ramey grapple feeds logs from the yard to the chop saw infeed slip. A 15-inch-diameter log is the maximum handled by the chipper-canter.

Top: Infeed line to the Beaver—looking from the rotary log separator back toward the debarker, the chop saw operator's station, the chop saw, and the grapple operator's station. Rotary log separator is reversible and can be used to kick oversize logs back into the yard if necessary. *Bottom:* A closer look at the rotary log separator and barked logs on the Beaver transfer deck. Eventual plan is to use rotary separator to sort logs to 26 inches in diameter.

Top: Elevated operator's station gives sawyer clear view of log transfer deck and infeed to the Beaver chipper-canter. Mirrors at outfeed to Beaver ensure view of cants. Scanners activate cant kickoff either to the resaw or to storage chain (in far left foreground). *Bottom:* A second Ramey grapple stacks surplus cants from the storage chain for later processing if flow is too heavy for the resaw.

debarker cuts the size to the maximum accommodated by the Beaver. After the log leaves the debarker, it moves along the conveying system to a stop which activates a rotary log separator. Eventually the log separator will be used to separate logs by size up to 26-inch diameters, to move those larger than the maximum diameter that can be handled by the Beaver in the opposite direction for a different kind of operation that can use the larger size more efficiently. Oversize logs are now traded or sold.

Chipper-canter is primary breakdown machine

Logs 15 inches and smaller then move on a transfer to the Stetson-Ross Beaver chipper-canter machine. The infeed mechanism consists of a rotating device (two reversible shafts equipped with sprockets that rotate the log to the proper position) and lift cylinders, which raise the log into position. At that point it is picked up by what we refer to as a buckhorn infeed, log gripping devices which feed the log forward into the chipping heads. The operation is controlled from an operator station behind and above the machine. This gives the operator full view of the incoming logs on the transfer and the logs entering the machine. Mirrors are positioned to allow the operator to view the cants as they leave the machine. The mirrors also show any improper machine performance as the wood moves through.

As the cant leaves the machine, a marking device indicates the entry position for the resaw as well as the modular size for optimum recovery. Optical scanners on the outfeed rollcase activate the cant kickoff mechanism. This, too, is reversible. One mechanism can transfer the cant to the resaw; the other, which moves in the opposite direction, can transfer the cant to a separate storage chain. If a surplus develops, a second Ramey grapple can stack the cants for future use in the event of a shortage of material to the resaw. This enables us to keep the flow of raw material to the resaw constant and smooths out production.

Infeed to resaw determined by operator

The transfer to the resaw is equipped with a pin-stop device so it can continue to move until it is filled. The pin stop is operated by the resawyer, who has the material moving toward him and can start, stop, or reverse it at will to keep a supply of cants moving on the merry-go-round system to and from the resaw. There are cant turners, angle rolls, and a Coleman-type roll infeed to the resaw. The vertical Coleman roll is activated by an Opticon electric eye device which moves in and out to various positions predetermined by the operator according to cant size. The sets of the resaw are also predetermined on a modular schedule controlled by diode cards in the electrical control system.

The resaw is a 5-foot Salem Equipment Co. twin band machine. The cants leaving the resaw move to an outfeed rollcase equipped with a sweep mechanism. The sweeps can be used either as an edging picker or as a cant separator to send the cants to the edger side or to transfer them in the other direction to a merry-go-round to go back through the resaw. Those cants which move to the edger are the nominal 2-inch cants, which go to a small two-saw edger, and 4-inch cants, which are bypassed by the operator to a

Sweep arms on resaw outfeed rollcase are edging pickers or cant separators. Cants go either to the edgers or to the merry-go-round (as shown), which sends them back through the resaw.

After the log enters the mill until it exits as lumber on the green chain, it need not be handled. A belt takes 2x4s directly from trimmer to planer. Other sizes go directly to sorting in rough or chipped surface condition.

second gang edger that converts them to a 2x4. The material which flows through the two-saw edger moves onto a simple 4-inch-wide belt. This narrow belt allows the edgings to drop from the side directly into a single-line by-product conveying system and carries the lumber onto the chain transfer moving to the trimmer. The same transfer that carries the lumber from the two-saw edger also picks up the lumber from the resaw and also from the gang edger as it moves from that point to an Irvington automatic gang trimmer.

Belt moves 2x4s directly from trimmer to planer

The trimmer has a memory system to separate the lumber destined for the planer; a timing device controlled by the trimmer operator activates a drop sorter mechanism. The 2x4s generally drop out to a transfer belt, which moves them to the planer. This is, we feel, a very effective arrangement. From the time the log moves into the mill until the lumber is ready to be pulled from the green chain, it is never handled by lift truck or carrier. The 2x4s leave the mill surfaced. Other thicknesses and widths go directly to the sorting chain in their rough or chipped surface condition.

If we want to run a size other than 2x4 (which makes up 75 to 80 percent of the production) the 2x4s can be separated at the trimmer to go to the green chain and the other size (1x4, 2x6, or whatever is coming out in any quantity) can be dropped onto the belt to the planer. We can then bring the 2x4s back during the regular run to smooth out the flow through the planer.

As lumber leaves the planer, it passes through an anti-stain treatment box by belt transfer to an extension of the green sorting chain, where the lumber is graded, stamped, and retrimmed if necessary. It then moves to the green chain, where it is sorted by length and grade and placed in uniform units ready for shipment.

By-product conveying essentially straight line

The by-product system reflects another series of tradeoffs in the effort to achieve simplicity in operation and maintenance. Since there are four principal by-products—chips, sawdust, shavings and bark—there are four essentially straight-line conveying systems

Vibrating conveyor separates sawdust from chippable material in front of the Morbark chipper. These chips join those from Beaver in the bin.

with a small number of laterals to make the system serve each machine.

The bark system is truly a single-line system—directly from the barker to a self-unloading truck, without benefit of hog or bin. It is the ultimate in simplicity, but not in efficiency, since it requires about three man-hours per day to haul the product away.

The chip system is a conventional high-pressure blower system; the chips from the Beaver enter through a rotary feeder at the blower end and the chips from the Morbark Chip-Pak enter the line through another rotary feeder farther down the line before the pipe begins to rise to the bin. The trims from the gang trimmer and from the planer trimmer, as well as the edgings and sawdust, move in a single system to a vibrating conveyor that separates the sawdust ahead of the chipper to another feeder and high-pressure blower to the bin. The fines from the chip screen also enter the sawdust feeder at this point through a gravity pipe. A lateral conveyor from the gang trimmer and the planer trimsaws also collects the sawdust from both edgers. A short 12-inch lateral auger serves the resaw.

Key positions have back-up operators

The operation was designed to function with one maintenance man. Occasions arise, however, when more than one situation requires attention at the same time. Some of the machine operators—particularly at the entering end, where capacity exceeds the capability from the resaw on—are involved in a portion of the regular maintenance work on the machines. Work schedules are staggered, to a degree, to accommodate the maintenance needs.

It was intended that each skilled position have at least one back-up operator. We have reached that goal. In most situations, the relief operator equals the productive capacity of the regular person and in some cases exceeds his performance. This might seem an insignificant matter, but its contribution to the continuity of productive capacity is inestimable. It required no capital investment that is measurable, either by us or by the tax collector, but we fully appreciate the value of the situation. Human relations do not respond to pushbutton control, but the simple application of interest motivation on the job pays substantial dividends both to the employee and to the employer.

We have attempted to assemble a system that utilizes mechanical principles for muscle applications and automatic-sequence operations, but we still rely on human intellect, skill, and judgment for those areas that offer a variety of challenges to reach maximum recovery goals.

I have indicated some of the things we feel contribute to the simplicity of our operation: (1) the simple spotting device used behind the resaw, (2) the log entry system, (3) the fact that we do not have to move lumber before it is in the surfaced condition to go on the car. It may take a few more men, but it required considerably less investment than it might have. We attempted to design this system in the most efficient manner with as little complexity as possible. This philosophy is based on a quotation I once read:

"Knowledge leads us from the simple to the complex. Wisdom leads us from the complex to the simple." We hope we have a reasonable quantity of each in the operation I have described.

DISCUSSION PERIOD

QUESTION: How many logs do you put through the Beaver per day?

SWANSON: The number varies. It is a difficult question to answer because we do not run the Beaver full time. We do not have the capacity to handle the material behind the Beaver, so we stack cants. The fellows working on the infeed end of the system may do something else for a period of two hours or more a day, while the rest of the mill is producing lumber. I really cannot give an average day's run.

QUESTION: Did you say the largest log processed is 15 inches?

SWANSON: The maximum diameter is 15 inches on the big end.

QUESTION: You do not have any log scanning equipment associated with the Beaver. Do you feel it is not necessary or that it is too complex?

SWANSON: We think we can train an operator to make a visual judgment on the logs and come out almost as well, considering the investment required for a sophisticated scanner to do the job. Our logs come in so many different shapes, lengths, and conditions that we think those judgments would have to be made even if we had a scanner-computer system. If we put in a scanner, we would have the operator there to override the scanner, which happens in many cases. It seems to us that it is a visual judgment in most cases anyway. You might be surprised how close our sawyer can come to the best opening face. Since our production is largely 2x4s, we are after about a 4-inch face on enough of the log to produce the product we want. If you get a log with a lot of sweep in it, you have to make some adjustments. You might have a log that would scale 6 inches, but you might set your top and bottom heads on 9 inches. Run that log through and you will get a face on the bottom and nothing on the top. If you do it some other way, however, you are running a lot of lumber up the blowpipe. So many visual judgments are required that we feel the operator can do the job just as well. I am not ready to give out any recovery figures, but I believe he is doing his job.

QUESTION: Did you say the piece count was 50,000 per day?

SWANSON: No, I said we produce about 60,000 board feet of lumber per day. Converting 60,000 board feet of lumber to a piece count would give a close approximation. The dollar end is the lumber end and that is what we are after. The piece count varies with the logs. Since we are a little bit out of balance with respect to log input and the capacity of our resaw to handle it, we balance it off by having the people on the infeed end do something else for part of the day. We do not run all the facilities full shift.

QUESTION: Do you put a continuous face on one side of the log and let the taper run out on the other side?

SWANSON: That is correct. It does not always come out that way, but that is the intent.

QUESTION: So you then put whatever face you can get on the sides? You do not clean them up completely.

SWANSON: Generally, the sweep on a log is all in one direction. Then, if you get less than a face on the top side, you get basically a flat side on the bottom. The sides will generally come out with a full-length face. You have to have a face somewhere to line up the cant at the resaw. There are ways to get around it, if the resawyer is on his toes, but it generally is not worth the extra time required to turn it and run it through the merry-go-round. There are some tradeoffs there in terms of economics. We try to train our people to understand what those tradeoffs are.

QUESTION: How many employees are involved at this mill?

SWANSON: It is programmed for 16 employees. We have 19 now, but we have some additional equipment to install to reach 16. We will get there. For instance, we now have one man stamping lumber, because we do not yet have a Sweed installed. As soon as we get that in, it will take care of one employee. We have another man driving the bark truck I mentioned. With a bin, we will take care of that. The operation is not quite where we expect it to go when it is completely finished.

QUESTION: You spoke of having back-up men for key positions. Are they also doing other jobs in the mill?

SWANSON: Yes, they are. We try to hire people qualified to handle more than one machine. We have a man on the green chain, for example, who is very versatile. We pay him a little bit more than green chain wages, but he is qualified to do other things and we just move him up the line. We have some men who can do two or three things, and we have some who can do one job in addition to the one for which they were hired. Most of the machine operators are trained to do more than one thing. It is just a matter of shifting people around. We do not have a whole group of extra people, but simply extra capabilities.

QUESTION: Do you have much of a turnover?

SWANSON: No, we have not had much of a turnover.

QUESTION: How is absenteeism?

SWANSON: Except for the flu, absenteeism has been very low. I might add that a large number of our people had only minimal exposure to a sawmill and a couple had never seen a sawmill before they joined us. They were better off because they did not have to unlearn some habits. We taught all these men from the beginning, while we were learning. It has worked out very well. We have discovered that most of it is attitude, rather than aptitude. If you get a man with the right attitude, it does not take long to teach him the rest.

QUESTION: Was this particular mill built from scratch or was it a conversion?

SWANSON: It was built from the ground up. We had another mill a short distance away where we did some of the operation while this mill was under construction. At that point, we were transferring lumber long distances. It certainly made us realize the improvement in not having to pick the lumber up, put it on a load, and move it back. Each time you pick it up, it costs money.

6

Attention to detail can reduce lost-time dollars

David H. King, President, King, Murphy & Associates Ltd., Richmond, British Columbia

The subject of sawmill problems can be broadly divided into three separate categories: (1) man management, (2) mechanical inefficiencies, and (3) recovery. I have chosen to limit this discussion to mechanical inefficiencies. Our participation in plant surveys in the United States and Canada prompted us to prepare this presentation.

You may consider some of the problems to be pinpointed unimportant. You may not have many of these problems in your operations. But I wish to point out all of these problem areas were important to the respective plants in which we identified them. We feel generally that plants can solve their problems, but sometimes it is beneficial to have an outside opinion. Often the problem appears to be unrecognized or, if recognized, it is not evaluated.

In recent years great improvements have been made in the design of sawmill production equipment, particularly for the processing of small- to medium-sized logs. But I will confine my discussion to general problems that exist in many sawmill operations. These general problems have great bearing on the profit and loss of an operation and are often overlooked or side-tracked because of larger single issues.

Most plants have wasteful practices that have become daily routine. These malpractices cause lost profit for three main reasons: (1) machinery shutdown, (2) manpower inefficiency, and (3) mismanufactured product. This, of course, is lost time, with a percentage loss in raw material.

The first approach is to define the problem and its effect on the operation. To do this, we must observe every part of the operation objectively. In industry, controls on expenditures make it difficult in many cases to correct faults. People therefore become complacent and learn to accept the problem as part of their everyday function. What are needed are measurement controls and a questioning attitude by supervision. For example:

1. How much production loss is caused by the problem?
2. What effect does it have on recovery?

One measurement control used by all operating plants is machine lost time slips. These can vary from sophisticated measuring meters that count actual sawing lines and record lost time to simple lost time slips filled in by the operator. If the lost time and the reason for the lost production is clearly defined, these slips become a valuable results measurement tool for the shift supervision and management. In many cases, supervision investigates the problems to determine how to eliminate this lost time, but, on the other hand, these slips are often read and then filed away.

I suggest that after the supervisor has completed his investigation, he then tabulate the amount of lost time under each heading that has caused the lost time. He then should request the mill accounting people to supply him with the cost per minute when the mill is not producing. This figure, of course, takes in operating labor, overhead, and loss in profit. With this figure, the supervisor can measure his loss in dollars. He should recap these results at the end of each month. After a few months, a pattern will develop, illustrating the costly problem areas. Now the supervisor not only knows where the problem areas are, he also knows how much money he can recommend to eliminate the dollar loss caused by a particular recurring item.

The tables throughout this chapter show the actual results for various sections of a mill where lost time for 1974 was carefully tabulated and defined. The cost per minute calculated by the mill's accountants for this single-line mill was $7.59 per minute. I will show in conjunction with these tables examples of some of the problems identified so that you may examine these areas in your plants.

To cover some of the problem areas in a sawmill, let us start at the log haul and work through the operation.

Improving operation in the log haul area

Table 6.1 lists various causes of lost time in the log haul and infeed areas—problems that amounted to more than $20,000 when viewed in cost per minute. Let us look at each of these areas more closely.

Those of you who have log hauls (infeeds) may find you lose time due to large spaces between logs, logs sliding back and downward from the curve, or logs cradling in the trough. To reduce lost time, some mills speed up their log hauls or infeeds. We recommend running the log haul as slowly as practical, but increasing the number of chairs to a maximum of nine links spacing per chair. If your log haul is steep, install chairs every seven links (Figure 6.1). This change can be made for a few thousand dollars. In many mills we have visited, spacing is 11 to 16 links between chairs. This is too large, and will cause the problems noted.

Chairs must have equal link spacing. Otherwise, the slip man feeding the haul has to do extra work. He pushes the log up to the

Table 6.1. Lost-Time Record: Log Infeed (cost - $7.59/min.)

	Hrs	Loss
Log hung up in haul	16.3	$7,400
Chain off sprocket	5.7	2,600
Loose wear strips	12.3	5,600
Apron jammed chain	6.3	2,800
Damaged cutoff saw	4.1	1,900
V-belts on saw	5.3	$2,400
Totals	50.0	$22,700

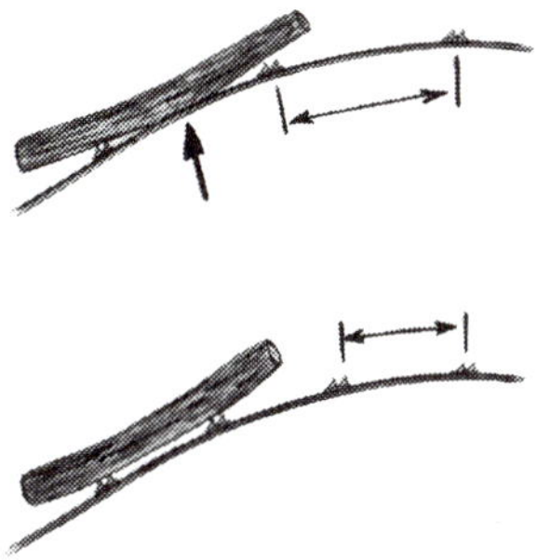

Figure 6.1

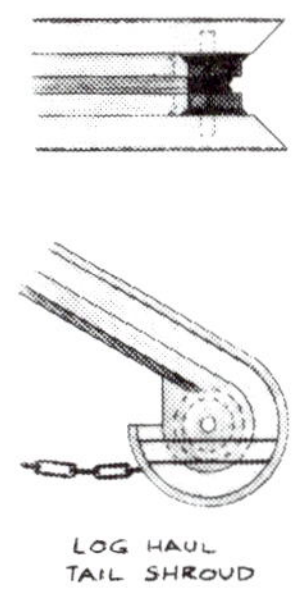

Figure 6.2

haul. If it is picked up immediately, he can go about getting the next log in; but if the space is too large or, worse still, a chair is missing, the log bounces back out. The slip man must then push it back in again. If the log haul chair is traveling too fast, it tends to kick the log away from the haul. If a large log has a short crook that fits between the widely spaced chairs, it will cradle in the trough. If a large crooked log does pull up to the curve, there is a possibility of the log sitting down on the wear strip, in which case it could start sliding back. This happens often and, of course, can damage the chain or even tear out the structure used for raising the apron.

Closer spacing of the chairs probably only means the addition of eight to ten chairs. These would be paid for very quickly with the lost time dollars. Some mills now use an articulating-boom crane mounted on a float to feed their log hauls. This, of course, is very positive. A good operator can feed a continuous flow with minimal space between logs.

We find other major contributors to lost time in the log haul area.

Tail drum shrouding. Improper shrouding around the tail drum will at some point allow the chain to be diverted and cause a hangup. This, of course, is costly in lost time, as the apron must be raised and the chain often must be cut to get it out from between the tail drum and the frame. Figure 6.2 illustrates proper log haul tail shrouding.

Wear strips. Over the years many types of metal have been used for log haul wear strips. Some choices were disastrous, as they were not compatible with the chairs, so chairs and wear strip wore out quickly. Today we have available some very reputable plastics that

1. Reduce friction and, therefore, also reduce horsepower requirements.
2. Reduce wear on the chairs.
3. Eliminate lost time due to wear strip tear-out.

Use 1-inch by 4-inch plastic strips (5 million molecular weight) to fight abrasion. This wear strip can be fastened down with a cartridge fastening system, whereby a shot-charged steel pin is fired through the plastic and the supporting steel.

Chain pitch. To prevent lost time caused by the chain jumping off the head sprocket, the sprocket teeth must be raised to match the elongated pitch of the wearing chain. In some cases the chain links will jam down onto the teeth (Figure 6.3) and, of course, this causes the chain to follow the sprocket around until the return chain tightens enough to pull it off the sprocket. This usually is the cause of the return chain jumping off the return rail.

Safety stops. In the event the log haul chain breaks or lengthens from wear, it is important that automatic stops be provided. In some cases, the chain has broken as it comes off the drive sprocket. The operator has no way of knowing this has happened, so he keeps operating the haul; consequently, he feeds the entire chain

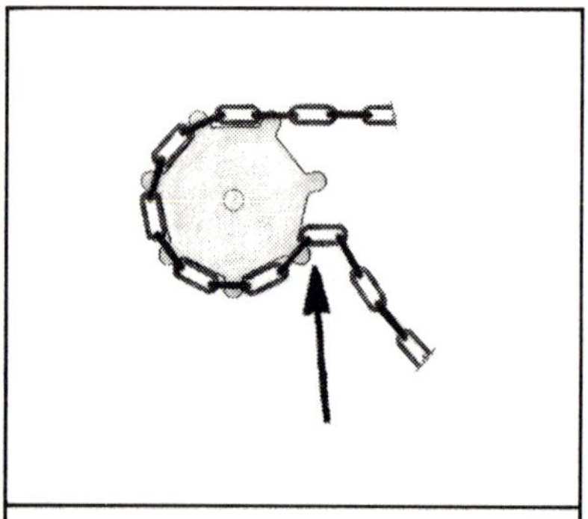

Figure 6.3

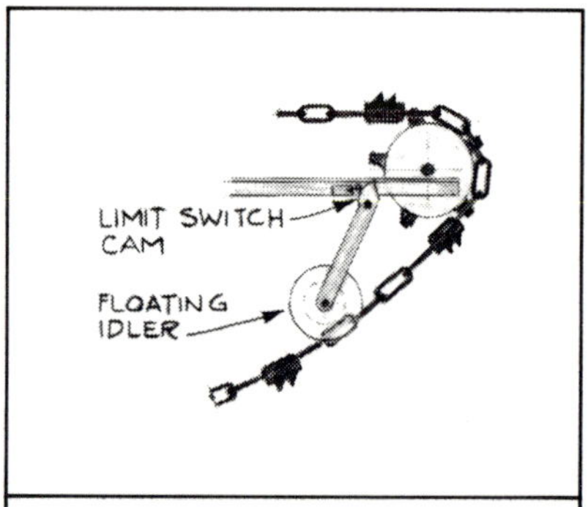

Figure 6.4

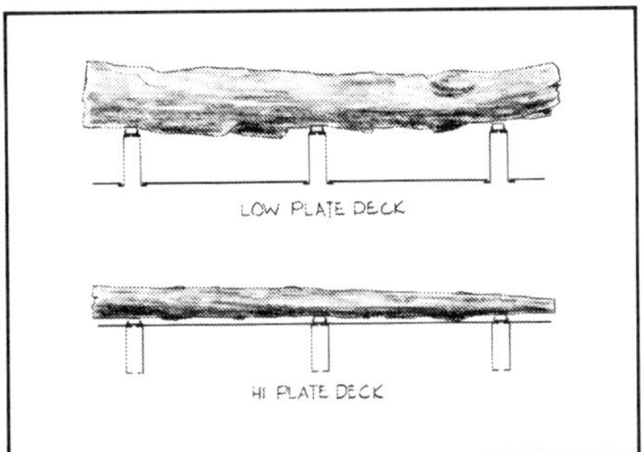

Figure 6.5

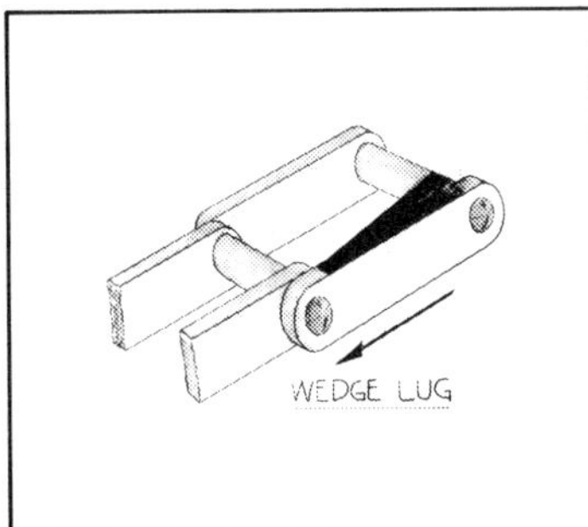

Figure 6.6

into the refuse conveyor. By installing a floating idler that will activate a cam switch, this problem will be eliminated (Figure 6.4). If the chain has lengthened through wear to a point where two links should be removed, the cam switch can be set so it will shut off the power to the log haul.

Reducing log transfer deck problems

As you are aware, there are many designs of log decks, either servicing waterborne logs or dry-land log handling. These range from sloping dead skids to log decks containing unscramblers and ending rolls.

Deck plates. When handling medium to large logs, it is necessary because of flare butts, large knots, or burls to keep the deck plates well below the top of the chain. This, however, is not the case when you are handling small logs. With small logs the deck plate can be placed on top of the stringers, which, of course, makes the distance below the top of the chain 2 to 3 inches. This will save problems with cleanup and with small, short logs turning and lying between the chainways. Figure 6.5 illustrates the two placements of deck plates.

Also notice that the sketches show the flanges turned inward. It is, unfortunately, quite common to find the flanges turned outward on most log decks. This causes logs to hook under them constantly and creates many operating problems.

On some decks we have had wedges welded on the chains (Figure 6.6). The sharp end of the wedge is always pointing forward so the logs can easily slip up and over if the chains are left running under the full deck of logs. The wedge lug will give enough pull to help the log clear the debris from the deck.

Deck drives. When handling long logs over decks that are fed with bundled logs from the water or from mobile wheel machines, it is a great advantage to have the deck headshaft split into two sections with a drive unit for each section. This will allow the operator to move either half of the deck to straighten the logs against the heel of the loader. Hang-ups on log decks cause a high percentage of the time lost in most operations. To minimize lost time, provide the operator with adequate equipment to enable him to cope with the problems. Additional capital expenditures certainly will be required to provide this equipment, but once installed, you will continue to reap the benefits.

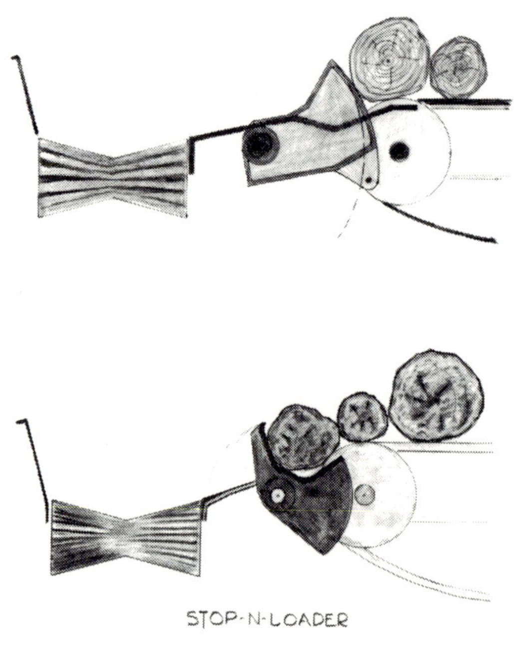

Figure 6.7

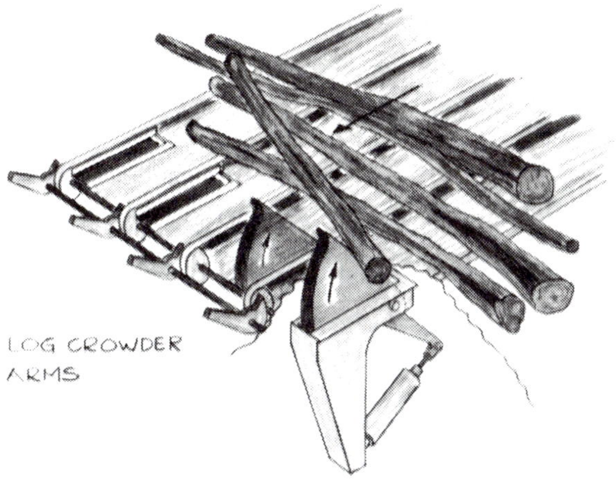

Figure 6.8

Log stop-and-loaders. Stop-and-loaders, as well as the log deck, should be split into two separate sections. The operator should have three control buttons, one for each section of loaders and one that he can use to operate both sections. The relationship between the loaders, the log deck, and the log conveyor is also important. In some operations the loaders are incorrectly sized and spaced. This is a serious enough problem, but even more serious, we often train our operators to put up with this poor design. There are, of course, other types of loader arms that you may wish to consider, but the two drawings in Figure 6.7 were done to illustrate the correct relationship between deck and roll case.

Crossed and tangled logs. On most log infeed decks, even those with waterfalls to a breakdown deck, you can have problems with crossed or tangled logs at the loaders. Some operations use an articulating-boom crane with a grapple to untangle crossed logs, as well as to load them into the log conveyor. As an alternative, we have developed a crowder arm design to cope with this problem.

The function of these crowder arms (Figure 6.8) is to crowd the logs back against the flow. This, of course, will straighten and untangle the pile. On long log decks the best method is to install four units, in pairs of two, at 12-foot centers. The first arm of the pairs should be approximately 8 feet from the end of the logs. The controls for the air cylinders are set up to allow the operator to operate either pair independently or both pairs together. If the logs are crossed and tangled on one end and they are coming up to the loaders with that end first, the operator will then only operate the pair of crowder arms on that end. He will throw the logs back, which, of course, will straighten and untangle the logs.

To avoid lost operating time, give the operator adequate equipment to do his job. When he does not have this equipment, he often gets himself into a mix-up that he cannot get out of. For example, I have seen operators lower one side of their loaders and move the crossed logs out over them in hopes of straightening them against the wall of the log haul. This usually results in two or more logs being forced into line and all falling into the receiving log haul together. The operator has made his problem worse and he cannot get out of it unless he can use cables to pull those logs out.

Log conveyors and fluted roll cases. Care must be exercised in the design of log conveyors or fluted roll cases to ensure that there will be a minimum of conveying problems.

Air pusher arms. One must never lose sight of the fact that logs will be sent to the mill incorrectly limbed and bucked. It is not unusual for an operator to be faced with loading a crooked log into a chain log conveyor. The log can have enough sweep so the ends will rest on one side of the conveyor while the center portion rests on the other side. In this case, the chain can slide under the log in the conveyor without touching the log. The experienced operator will have learned a few tricks to cope with this problem. Sometimes these can be effective, but other times they will hold up the log flow long enough to cause lost operating time. If you are experiencing this problem in your operation, you might consider installing air-operated pusher arms on each side of the conveyor (Figure 6.9). This equipment will allow the operator to nudge the log from either side to work the crook down onto the chain. Again, these arms are not required if logs are bucked properly before arriving at the mill deck.

Some mill operators prefer fluted roll cases to log conveyors.

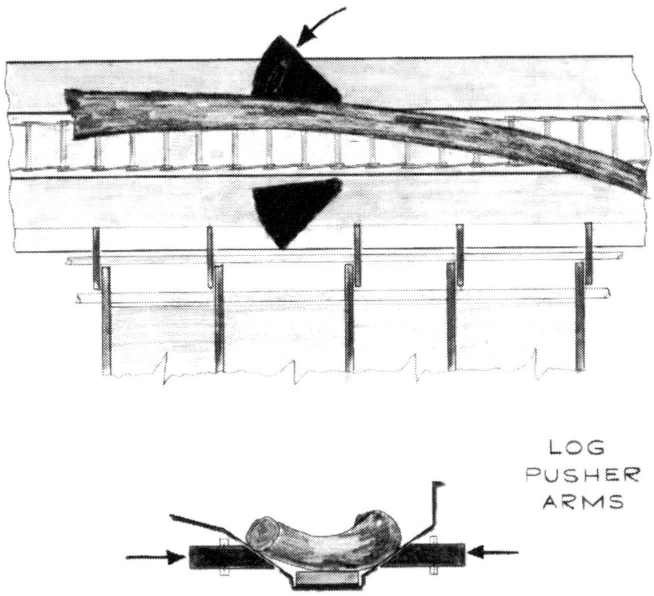

Figure 6.9

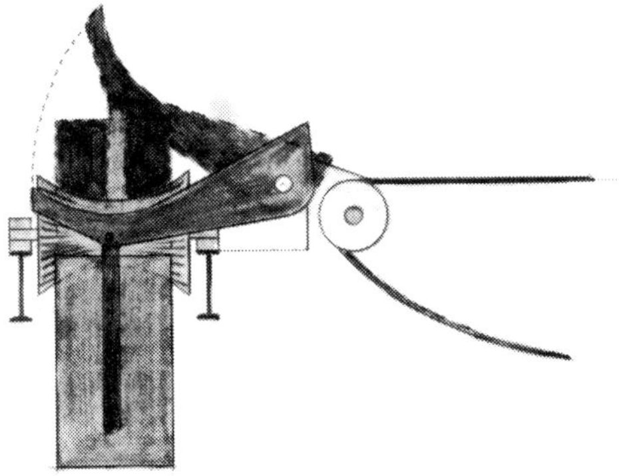

Figure 6.10

There are arguments for both. In either case, however, the log supply has to be studied and facilities designed to handle its particular problems. For example, if your log supply contains many rough logs with flared butts, pistol grips, and/or large burls and knots, then the standard fluted roll case can be very troublesome. If, on the other hand, you evaluate your log supply and install your fluted roll case with closely spaced rolls, you will find it will perform better than the chain log conveyor.

Belt conveyors have been used quite extensively for conveying short pulp logs. There are instances where belt conveyors can handle long logs, but it will be necessary to improve design standards to extend belt life.

Log unloader arms. When installing kickers or unloader arms from log or timber roll cases, we recommend you *not* install the lift-arm type. These are very troublesome, with pieces running under them while the operator is unloading a previous piece. In two operations that were unfortunate enough to have these, we installed a heavy steel plate attached to the first arm (Figure 6.10).

There are, of course, other types of loaders, such as top-mounted pusher blades (Figure 6.11, top), star unloaders, punch bar kickers, and sweep arms (Figure 6.11, bottom). There are some problems with all types, but these problems can be reduced by design features. It is always desirable, in unloader design, that you not have to wait for the arms to return. Random-length logs present a problem, as it is possible to unload a short log and also the front end of the log following. To prevent this, the kickers or unloaders should have two separate sections. The first section would have enough arms to handle the short logs; the other section, with separate controls, would have one or two arms to assist in unloading the longer logs. The operator would be provided with one button for short logs and one for long logs. If you are having problems in this area, you may wish to consider the under-roll rotating-sweep type.

Side lifts. Multiple-chain side lifts, if not correctly designed, can be very costly in lost time. They, like log hauls, have a few particular areas that cause a high percentage of the lost time. One of these is log chairs. The chairs can frequently get out of alignment, which, of course, can cause the logs to tumble back or spin around as they go over the head sprockets. The misalignment

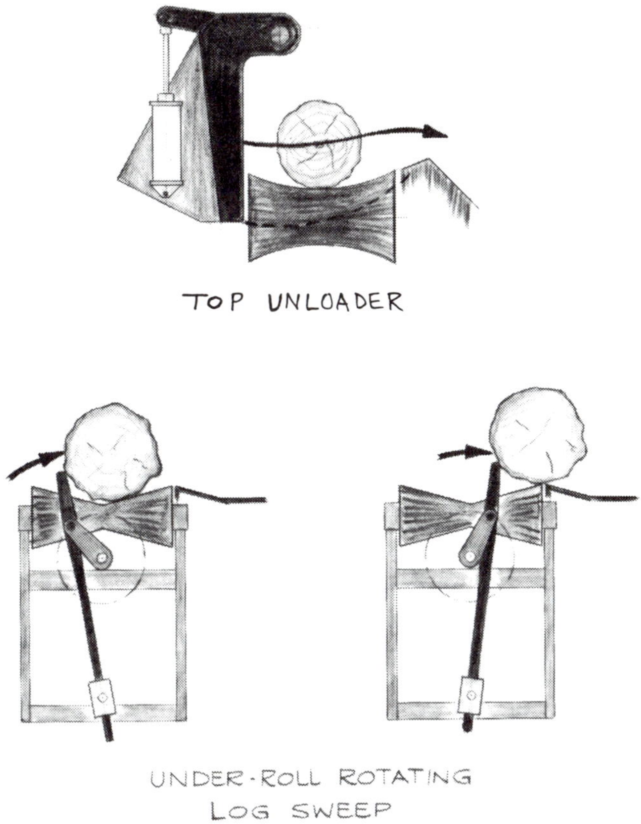

Figure 6.11

problem is usually caused by failure of the design to ensure that logs are cleared away from the delivery point; in this case, one or two sets of chairs may get caught and cause the chains to jump on the drive sprockets. In most plants the cutoff saw operator controls the side lift operation; he also has many other functions to perform. On occasions when he is fighting a hang-up in another area, he might allow the side-lift delivery transfer to pile up. To eliminate this hang-up, there should be an electric control that will stop the side lift when the delivery transfer is loaded.

To assist the maintenance staff in realigning these chairs, it is important that the side lift be designed in a way that will reduce this to a simple task. Many hand injuries have been sustained when millwrights and/or operators tried to lift these chairs back into line. We recommend a design which uses a jack shaft drive; each chain is then connected to it with a roller chain drive (Figure 6.12). With this equipment, aligning the chairs is just a matter of removing the roller chain drive on the affected chain run and then jogging the remainder until they are all in line.

If the side lift is fed from the water, it is important to enclose the return chain from a point above the high water, extending down around the tail sprocket and up the front a foot or so. This will prevent deadheads or chunks of log from getting caught in the return chairs. On side lifts operating from a dry deck, it is important to provide proper chuting and a cleanup conveyor to keep debris clear of the return chains.

Chain tighteners. To prevent the chairs from pulling the chains out from the face of the side lift when loaded, it is important to use a heavy, floating tightener on each of the return chains.

Bundle log lifts. Many operations have changed from flat rafts to bundle booms. When feeding logs to a conventional log haul or

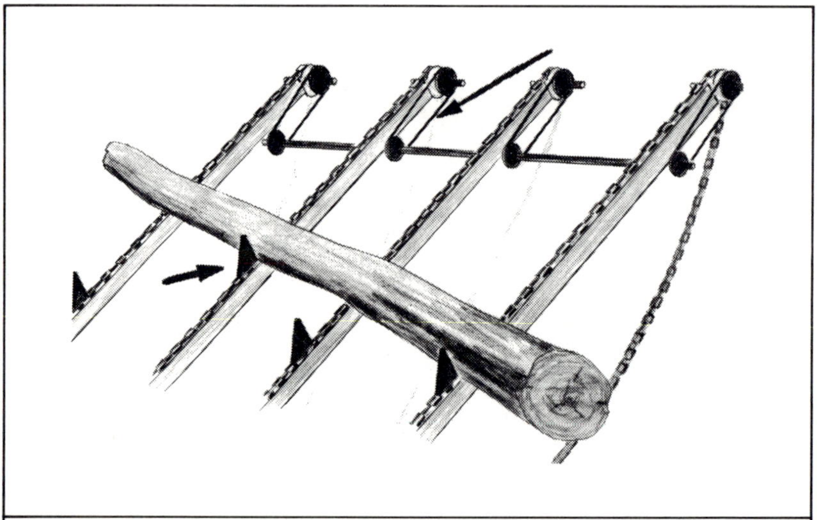

Figure 6.12

Table 6.2. Lost-Time Record: Debarker (cost = $7.59/min.)

	Hrs	Loss
Broken hoses	7.7	$3,500
Electric controls	5.4	2,500
Infeed and outfeed chains	46.5	21,200
Oversize log jam	16.2	7,400
Damaged abraders	12.3	5,600
Maintenance	17.7	8,000
Electrical—power	7.7	$3,300
Totals	113.5	$51,500

side lift, it is necessary to open the bundle in the water and feed out one log at a time. To avoid this wasted work, equipment has been designed to lift or convey the bundles up onto log decks for breakdown.

Determining debarking inefficiencies

I will only touch lightly on debarking as a thorough discussion of the many variables would take more than the space allotted to the entire discussion of sawmill problems. Table 6.2 points out the lost-time cost of some problems at the debarker. The three main problem areas that usually contribute to high operating costs are feeding, centering, and abrader arm maintenance.

Debarker feeding. To minimize time lost at the debarker through logs hung up at the infeed, have well-designed and well-manufactured equipment feed into and out of the debarker. It is very important that both infeed and outfeed be adequate. Do not skimp in these areas. If a chain infeed is used, be certain that it will handle misshapen logs, not just straight logs. The chain and chainways must be designed so that crooked logs cannot crowd the chain off the sprockets.

Fluted rolls can be used effectively, providing the rolls are closely spaced. When using small-diameter rolls, they should be spaced at 2 feet 6 inches center to center, whereas the larger diameter rolls would be 3 feet center to center.

Log or ring centering. When handling large logs, in most debarking operations the debarker ring is centered to suit the log, whereas the log is usually centered to suit the debarker when handling small logs. To achieve optimum debarking and minimum debarker maintenance, it is absolutely necessary that the log pass through the abrader arms as close to center as possible. Figure 6.13 illustrates rubber-wheel centering for both the infeed and the outfeed. There are other types of equipment available. *All* debarkers should be equipped with a centering device.

Abrader arm maintenance. To attain the best possible barking with a minimum of abrader arm pressure, abrader tips must be kept sharp. There is every reason why the abrader arms should have attention equal to that given band or circular saws. Maintaining the correct self-opening curvature of the arms is important to give simultaneous smooth opening of the abraders.

The main design problem with mechanical debarking is the thrust required to debark. If two rings were used—one rotating clockwise and one counterclockwise—thrust would be neutralized. Thus a narrower abrader could be used, due to the increased coverage provided by having two rings. The reduction in thrust would remove the twisting action that breaks many logs, particularly small logs, in the winter. The additional ring would increase the capital cost of the debarker substantially, but I would be willing to predict that the difference in maintenance, plus the improved debarking, would pay for it in less than five years.

Log cutoff saws. Log cutoff saws are usually a problem when bucking small-diameter long logs. Many mills install a saw on the log conveyor and pass the logs by the saw. This is very slow, as the

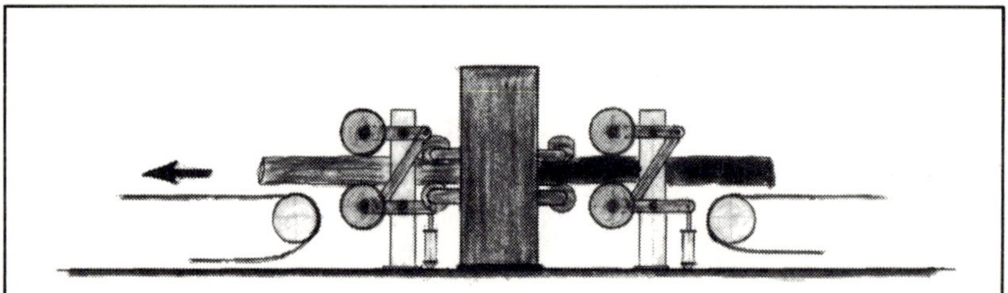

Figure 6.13

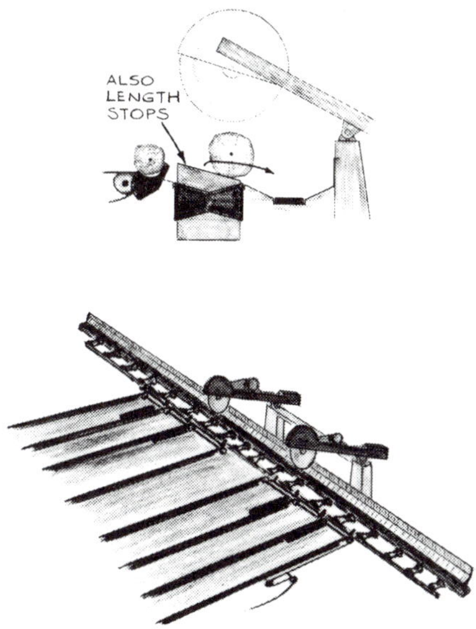

Figure 6.14

saw sits idle while the log is conveyed into position. To insure an adequate log supply, consider a more productive operation.

One particularly good method (Figure 6.14) is to have a fluted roll case, a chain log conveyor, and two chop saws. The log is loaded onto the fluted roll case, which in turn positions it for correct bucking. The two chop saws reach across the chain log conveyor and cut the log on the fluted roll case. The space between the two cutoff saws is determined by the popular length of log wanted in the mill. Variations require only a short ending travel. After the log is cut, the pieces are loaded into the chain conveyor for delivery into the mill. Figure 6.14 illustrates three additional design elements:

1. Log deck headshaft is split into two sections with a separate drive for each section.
2. Loaders are split into two sections.
3. Two sets of log crowder arms are provided.

A look at the key area—the headrig

In the many operations, the sawyer and his equipment have not received adequate attention. The sawyer must make so many actions and decisions that it is necessary for him to develop a rhythm. He has to acquire skills similar to those of a pianist to enable him to reach a high production level. To enable the sawyer to attain this high degree of skill, we must give him equipment that is designed correctly. He cannot have a piece of equipment that operates correctly only part of the time. This upsets his rhythm and requires that he start all over again—hence frustration and low production. If you give a stenographer a typewriter with some keys that malfunction, you can imagine what kind of a result you will get. The following are problem areas worth examination.

Sawyer's cab. It grew to be the practice many years ago to situate the sawyer at mill floor level, between the bandmill and the log deck. This was wrong then and is wrong now. The man who determines the production level and the quality of production thus has the most awkward operating station in the mill. With the advent of chipperheads in front of the bandmill, it became necessary to raise the sawyer up above the chipperhead. Initially this move caused some experienced sawmill people real concern. But

Figure 6.15

when questioned later, they stated they would never again lower the sawyer's cab, even if they had no chipperhead.

The major advantages of having the sawyer in the raised position (Figure 6.15) are three:

1. *Log loading.* While he is sawing one log, the sawyer can glance out over his log deck and learn the characteristics of the next log. This saves him time, as he knows exactly what action is required to load the next log. When he knows the diameter of the log, he can move the knees on the return trip so they are correctly positioned to receive the log. In his previous location he could see only one end of the log.
2. *Log turning.* In the raised position, the sawyer can turn the log as soon as he is clear of the saw. He can also do this in the lower position, but he has to be very careful or he may throw the end of the log into his operating station. There was a case about a year ago where a sawyer was cutting a hollow log which split into three or four pieces as it was moving toward the bandsaw. The sawyer brought his turning bar up to hold the broken pieces onto the carriage. He misjudged and hooked one piece, lifting it high enough so it entered into his operating cab. It pinned him against the wall, killing him instantly. Had he been operating from an elevated position, he would have been out of the line of fire. It should also be noted that in this raised position he is safe when a saw breaks.
3. *Sawyer's visibility.* It is an important part of a sawyer's function to see the tail-sawyer's station and the roll case away from the carriage. To do this, he must continually lean forward to see past the husk. With the cab raised, the sawyer can see through the spokes of the wheel. He can also see the operating parts of carriage more clearly. Many operators fear the sawyer will not be able to see his log to grade saw properly. I am sure some sawyers, without examining an actual installation, would have the same opinion.

Cabs should be completely glassed in and have heating and air-conditioning. In the new cab, the sawyer will not be subjected to dirt and sawdust. This modification has so many desirable features that it should not be rejected without watching an actual operation. When observing, visualize the cab customized to suit your particular requirements.

Shadow lines. Many mills do not provide the sawyer with a good shadow line. This can cost in recovery, as the sawyer is really guessing where his first line is. If he guesses wrong, he can saw a slab 1/2 to 1 inch too heavy, or he can just skin the log and waste a trip. Considerable money is spent to reduce kerf. This is desirable, but shouldn't a shadow line be considered first?

Bandsaw lubrication. We recommend removing the water from the guides on bandsaws and replacing it with an oil mist spray. It may be necessary to experiment to find the correct oil mix. On some applications on the Coast, we have used 70 percent hydraulic oil and 30 percent diesel oil. The advantages were:

1. Rotting of bandmill support timbers was stopped.
2. Sawdust does not stick to the bottom housing and build up around the wheel.
3. Freeze-up problems during the winter months are eliminated.
4. The life of the guide plugs is increased.

A bandmill lubrication system is illustrated in Figure 6.16.

Log loaders. Whether you have a single cut or a double cut, it is important that the log deck and the loaders are set at the correct height to provide sufficient slope to the carriage, as shown in Figure 6.17. It is usual to observe either of two extremes. Some mills operate with flat dead skids between the loader and the carriage. Others operate with too steep an angle. The flat dead skids are slow for the sawyer, as he wants the log to roll out of the loaders fast enough so he can anticipate the time required and at what position his carriage should be before he trips the loaders. He is required on many occasions to use the log turning bar to assist the log across the dead skids. This breaks his rhythm and slows down production. The steep angle, on the other hand, allows the sawyer to load quickly, but constant heavy impact necessitates extremely frequent carriage maintenance.

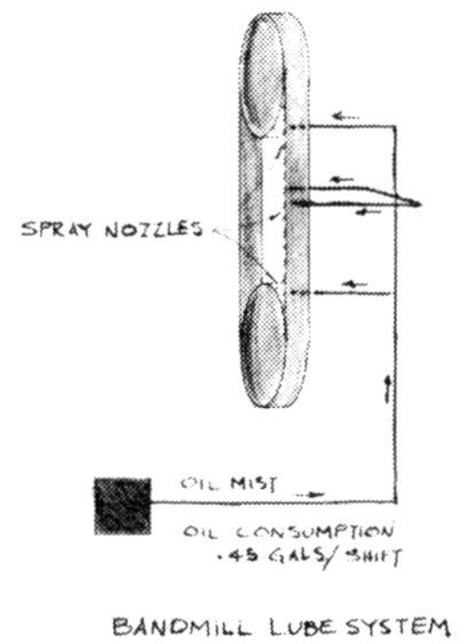

Figure 6.16

Figure 6.17

If I were operating a 40-foot carriage and my log supply contained many poorly shaped logs, I would consider splitting my loaders into two separate sections. Each section would have independent controls, thus allowing the sawyer to get out of trouble when the log crosses in the loaders.

Log turning bar position. When determining the position of the log turning bar, it is important to examine three major considerations: (1) the length of the logs, (2) the position of the knees on the carriage, and (3) the relationship between the log turning bar and the carriage knee when the log is clear of the saw ready for turning.

In many cases the carriage knee spacing is not considered when locating the log turning bar (Figure 6.18, top sketch). This oversight often places a carriage bunk in front of the log turning bar when the log is in position to turn. This requires that the sawyer worry about this obstruction and, of course, often causes bent bars, as well as damaged slides on the carriage bunk. The bottom sketch in Figure 6.18 illustrates a better placement.

Carriage bunk. Figure 6.19 illustrates a bunk design which effectively shears the log turning bar back to avoid hooking the side of the bunk. It also has a protection plate that extends up to the top of the bunk wear plates to prevent the log turning bar from bending the wear strips. Considerable offsized lumber is

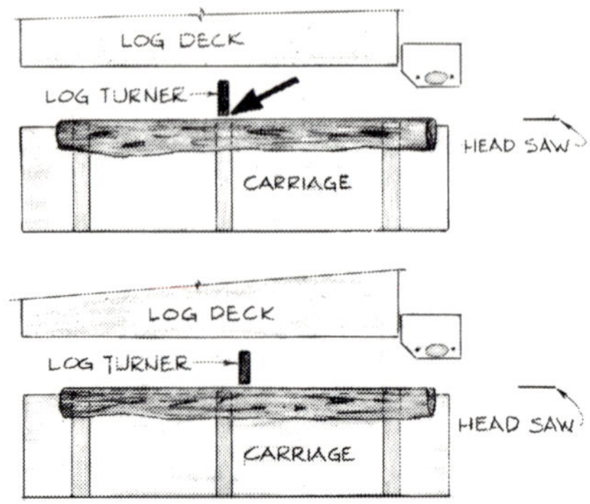

Figure 6.18

produced when the cant is held up off the bunk by the bent ends of the wear strips.

Knee cant turners. The knee cant turner, commonly known as a Brownsville turner, has been a popular addition to carriages for many years. The problem area we have observed in many of these installations is that these turners are installed in mills where the log deck is equipped only with a log turning bar. For effective use of these knee turners, we recommend installation of a rugged, fast set of push, or reload, arms (Figure 6.20). The sawyer flips the cant a quarter turn with the knee turners and then quickly pushes it back to the knees with the push arms. After he becomes accustomed to using these push arms, he will find he may use the

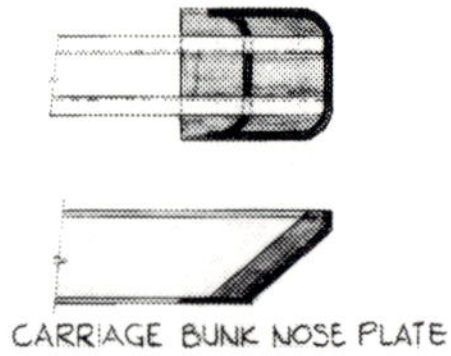

Figure 6.19

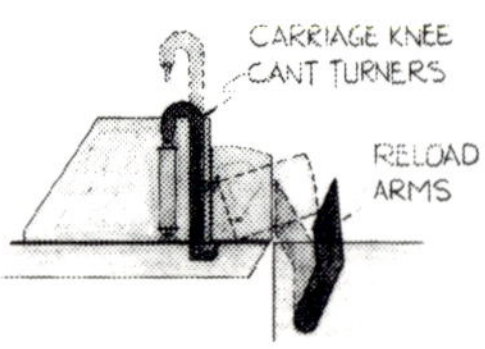

Figure 6.20

log turning bar only for loading and sometimes for his first turn. A sawyer who has never used push arms will require some convincing and retraining to accept and achieve this potential for increased turning and handling speed.

How saw speeds affect performance

A really significant problem area in many sawmills is the peripheral speed of circular and bandsaws. There is a tendency in the industry to run the saws too fast. Previous Sawmill Clinic volumes include some good information regarding saw speeds.

Some years ago there was a popular belief that circular and bandsaws should operate at 10,000 feet per minute. This rule of thumb was used by many operations, without recognition of the problems it created. To attain the ultimate in sawing, it is necessary to determine the specifications of the saw after considering the material to be cut, the feed speed, and the peripheral speed. Here is an actual situation:

Peripheral speed (9-ft bandmill)	10,000 fpm
Saw blade split gauge	0.089
Tooth space	1-7/8 in.
Swage	0.175
Difference	0.086
Carriage speed average	250 fpm
Size of chip kernel	0.046

The sawdust kernels were so small that they were sliding out of the gullet and packing next to the saw blade. (The size of the chip was 0.046 inch. Half the difference between the swage and the tooth thickness was 0.043.) This sawyer's production was restricted. If he fed too fast, the saw would produce snaky lumber. Our recommendation to correct his problem was this:

Peripheral speed	8,500 fpm
Saw blade (12 gauge)	0.109
Tooth space	2½ in.
Swage (6 gauge)	0.203
Difference	0.094
Carriage speed average	300 fpm
Size of chip kernel	0.122

Figure 6.21 shows the ultimate in sawing. The gullets are full of sawdust and on the right we show a clean saw coming down through the cut with plenty of clearance on the side. In actual operation this is never attained, but through correct specification we must try to achieve the best conditions for the saw.

Peripheral speeds and feed speeds of saws is a very complex subject. It requires a separate session to cover even the preliminary factors. Many years ago I spent considerable time, with an inventor from Spokane, in development of teeth and saws to cut chips when sawing lumber. We designed and patented teeth for both circular saws and bandsaws. We achieved good results, but the project was dropped when it was decided that the sawdust would be used for groundwood in its present form. During the six years we were involved in this project, we learned something more about what happens with a saw when cutting wood.

Most important, you must allow the tooth to bite into the wood, cutting a good-sized kernel. When cutting the chips, we reached in and cut as much as 1/2 inch with edger saws and 3/8 inch with bandsaws. Finally the chip length was set at 5/16 inch. I am not recommending anything as ambitious as this; I merely mention it to give an idea of the outer limits.

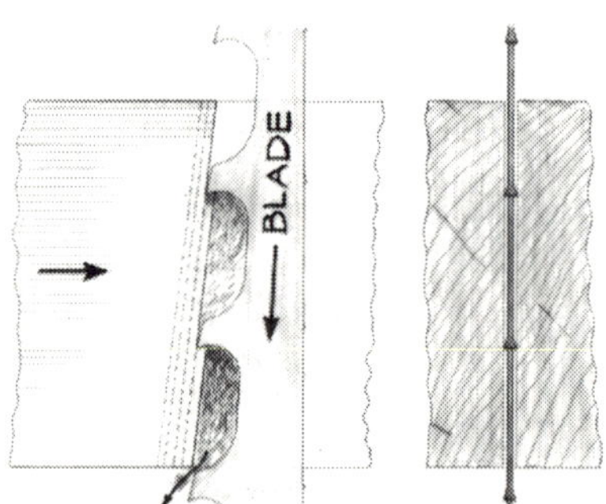

Figure 6.21

In a high-production bandmill, I consider a chip length of 1/8 inch to be realistic. A percentage of the sawdust from this saw will remain in the gullet. How much is a question we cannot answer, as we do not know how thick the chip is after it has entered the gullet. In the past, we recommended a chip length in excess of the clearance on one side of the saw. This certainly functions better than a saw that is cutting a chip that is less than this clearance, but still not at the optimum. The problem is the separation of the growth rings and the thickness of the growth ring. It can be a thin wafer, which will also turn and slide out of the gullet. We now consider it important to reach in and cut a longer chip that is also more than one growth ring in thickness.

The journey through the cut for the chip that is contained in the gullet is a rough one. The chip is constantly abused by the roughness of the cant walls. The heavy chip that comes off the point of the tooth is finally discharged at the bottom of the cut in a smaller form, due to this abuse. If you want to observe the difference, have the sawyer set his log or cant out into the cut 1/4 inch and then feed it past the saw. The chips that are discharged will be nice square kernels. Now have the sawyer cut off a side plank. Compare the two samples of sawdust. The sawdust from the buried saw is separated at the growth rings. Usually the thin wafers are the winter wood and the woolly sawdust is the softer summer wood.

We found it necessary when cutting 5/16-inch chips to design our saw with cleaning teeth between the chipping teeth. These teeth were so designed that they sheared the broken fibers off the cant wall, thereby smoothing out the journey for the heavy chip.

Figure 6.22 (top) explains what we feel happens when a band-saw tooth is not cutting enough wood. The arrow on the right shows the tooth trying to cut; but it is not cutting, it is rubbing. When the tooth is only cutting a few thousands, the effect will vary with different species, but generally we feel that the wood is being compressed ahead of the tooth, causing this rubbing action on the point. Cedar is a good example. If you do not reach in and cut a good sized bite, you have a difficult time with wood compression and shattering. The sawdust produced is so fine that it cannot be contained in the gullet. The rubbing on the point of the tooth creates heat and dulls the cutting edge.

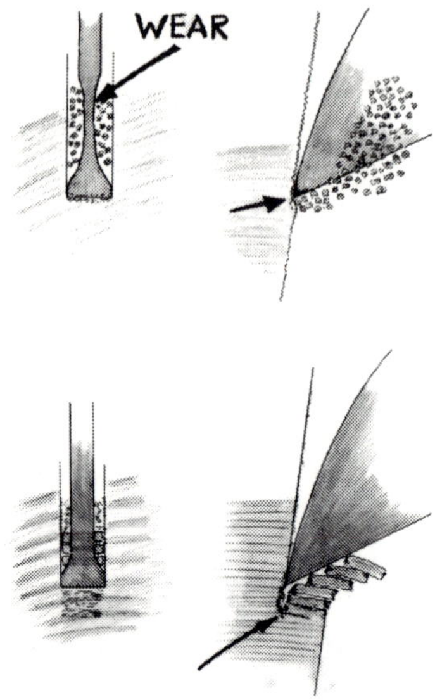

Figure 6.22

Rather, the tooth action should look as in the bottom illustration in Figure 6.22. It should be reaching in and breaking the fiber away before the point of the tooth ever gets there. The wear on the top of the tooth from the fiber in the cant is offset to some extent by the wear of the chips sliding underneath the point. You are really getting a self-sharpening edge. The chip in this case is quite large—1/8 inch in length.

The angle of the annual growth rings, relative to the saw path, also has some detrimental effects on the saw when cutting a small chip. We contend that the velocity of the saw, together with this grain angle, drives these small chips to one side of the saw. This packing of chips forces the saw out of line, thereby causing it to snake in the cut. This is particularly so after you have taken two or three cuts off one side of the log and then turned it down; now the majority of the grain slopes to the outside. The larger chip overcomes this problem, as the gullet carries it and only a percentage slips by to the side. When edger saws were subjected to excessive speeds, it was found that the points could not stand the heat; therefore carbide tips were born. The illustration at the top left of Figure 6.22 shows the wear that occurs when the sawdust is sliding out of the gullet.

Eliminating trouble spots at the edger

There has been considerable new development with different types of edgers over the past few years. Because of these developments, edgers in most of the mills we have surveyed have been reasonably efficient. For that reason, I will not discuss the edger to the same degree as the headrig. But I have picked out a few items of interest.

Edger spotting chains. When processing the long waney-edge side cuts from the headrig, it is common practice to have the edger spotting chains split. This is done to allow the operator to position either end, to obtain maximum recovery when sawing this material live. In 20- to 24-foot mills, there seems to be a trend to run the live lumber against a straight edge. This may not be too wasteful with timber that has very little taper, but I have seen this done in mills where it caused considerable loss, as well as poor slope of grain.

Edging tailing bars. There are many designs of edger tailing bars. Most designs work well until you are faced with timber that has so much bind that it closes the cut immediately behind the saw. Extend the tailing bar right up to the back of the saw. Figure 6.23

Figure 6.23

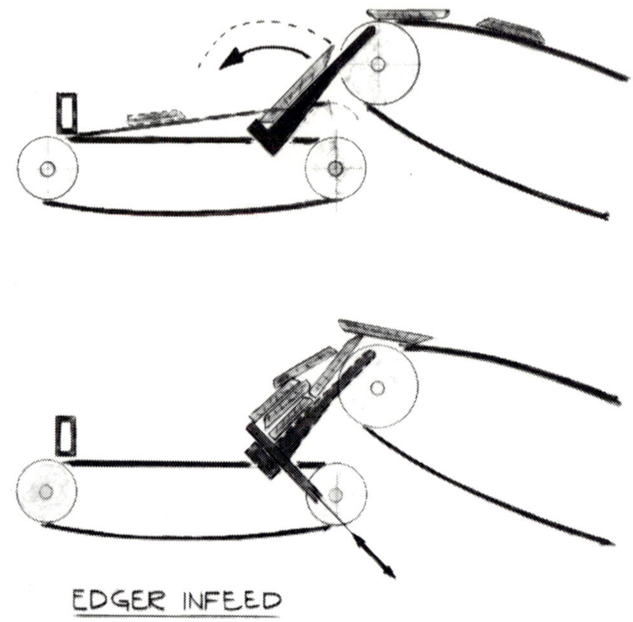

Figure 6.24

illustrates how we enclose the feed rolls to stiffen the shallow portion directly under the roll. To reduce the friction of moving this tailing bar, we support the slide bearing on a revolving shaft.

Edger feed tables. Some dimension mills have a problem obtaining a high piece count when faced with a mixed flow of lumber and waney-edge material. There are many transfer designs; these usually are tailored to suit the particular mill. Figure 6.24 (top) illustrates a design we use quite extensively. Lumber slides downward from the headshaft onto the sloping turning arms. If the lumber requires turning, the operator merely flips it over with the arms toward the straight edge. If it does not require turning, he raises the live chain skids and picks the lumber off the turning bar.

On occasions lumber will come to the operator doubled up and will give him a problem when two or three pieces slide down from the headshaft. To combat this, we install a set of pin stops at the base of this slide to hold back the extra pieces (Figure 6.24, bottom). Now the operator has to manhandle these pieces off the pin stops until he reduces the jam to a single piece.

Feed control. Recently there appears to be increased interest in methods to control the feed speed of edgers automatically. This is a problem in cant edgers as well as in gang edgers. One good approach is to install a direct current drive on the feed rolls and control this from the ampere draw of the arbor motor. This is also beneficial for planers where the top head motors signal the direct current drive. An example is provided in Figure 6.25.

SAMPLE DOUBLE ARBOUR EDGER

Motor	1750 R.P.M.
Motor pulley	12.5
Arbour pulley	14
Saws 16 tooth 9 guage	24" dia.
Arbour speed	1562.5 R.P.M.
Chips per min.	25,000
Feed speed	90 F.P.M.
Chip length	.043

CONVENTIONAL

Motor	1750
Motor pulley	14
Arbour pulley	13.3
Saws 18 tooth 24" dia.	
Arbour speed	1842
Feed speed	90 F.P.M.
Chips per min.	33,156
Chip length	.032

Clearance on each side of the saw is .035 to .040

Figure 6.25

Trimmers, unscramblers, and resaws

Feeder's position. In some mills the transfer conveying the lumber up to the trimmer feeder is too low (30 or 32 inches), requiring the operator to work stooped over all day. The height of the transfer should be about 36 inches. In addition, as he feeds lumber down onto the trimmer feed chains, his hand is exposed to the zero chain and lugs. Figure 6.26 illustrates one solution: the zero chain tail sprocket is moved ahead enough to remove exactly one lug space of chain. This may not seem significant to us, but it is to the trimmer feeder.

Saw settings. It may surprise you to know that many selective trimmers are operating with incorrectly spaced saws. Usually an operator states what overlength he wants, starting with an 8-foot length, and then each saw will increase in length by the thickness of the kerf. This allows the trimmerman to saw any given length on any position on the trimmer and get the same sawn length as he would if he cut it from the zero saw.

Figure 6.27 shows the actual saw spacing at one mill surveyed (top) as well as what that spacing should have been (bottom). Note how many short boards were being cut. This is particularly wasteful in the planer mill where the wand ahead of the closest saw was not touched, so the next saw automatically drops, cutting

Figure 6.26

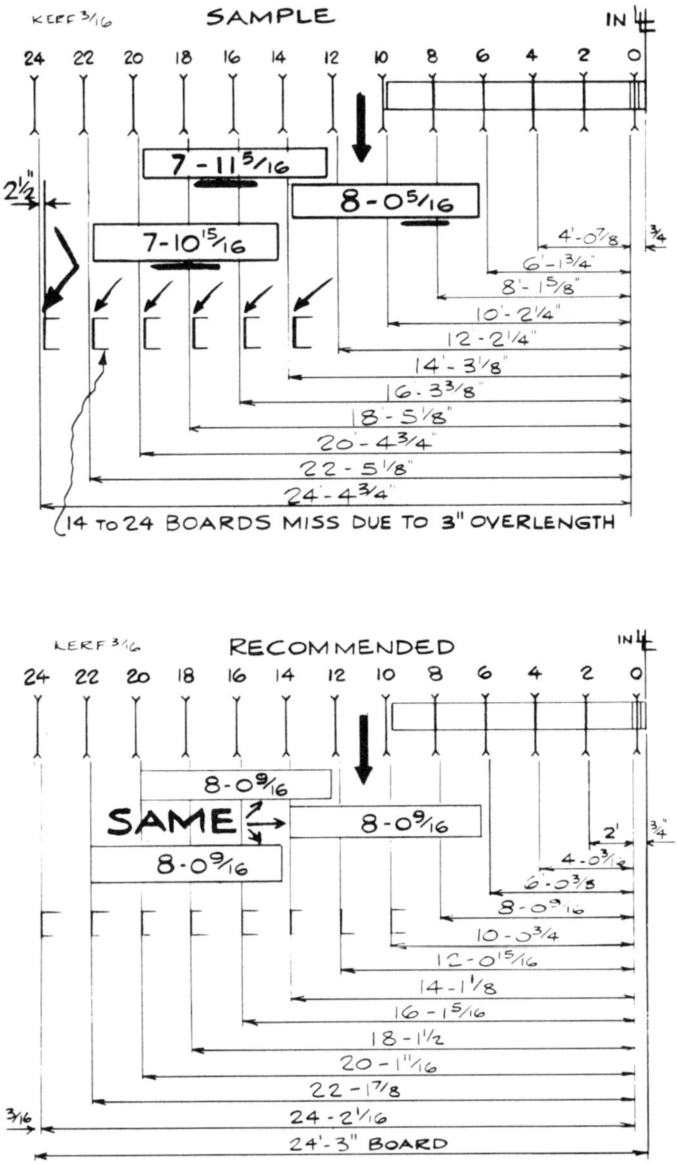
KERF 3/16
SAMPLE
IN
24 22 20 18 16 14 12 10 8 6 4 2 0
7 - 11 5/16
8 - 0 5/16
2 1/2"
7 - 10 15/16
4'-0 7/8
3/4
6'-1 3/4"
8'-1 5/8"
10'-2 1/4"
12 - 2 1/4"
14'-3 1/8"
16-3 3/8"
18'-5 1/8"
20'-4 3/4"
22 - 5 1/8"
24'-4 3/4"
14 TO 24 BOARDS MISS DUE TO 3" OVERLENGTH
KERF 3/16
RECOMMENDED
IN
24 22 20 18 16 14 12 10 8 6 4 2 0
8-0 9/16
SAME
8-0 9/16
8-0 9/16
2'
3/4"
4-0 3/16
6'-0 3/8
8-0 9/16
10-0 3/4
12-0 15/16
14-1 1/8
16-1 5/16
18-1 1/2
20-1 11/16
22-1 7/8
3/16
24-2 1/16
24'-3" BOARD

Figure 6.27

off from 23 to 24 inches of trim. How many trimmers now in sawmills have any adjustment for spacing the saws? When purchased, your trimmer may have had the correct spacing, but as maintenance work is done and ladders are replaced, the spacing is disturbed. A considerable trim loss can be attributed to this problem and the operator is often not aware of it. The sawmill trimmer is only one step in the overall operation from woods bucking to cutoff saw, to sawmill, to planer mill. Incorrect length selections at any one of these stations can cause waste throughout.

Unscramblers. Known to some as a board dealer, to others this piece of equipment is an unscrambler. I like the term *board dealer*, as I feel certain whoever conceived this idea originally meant it to

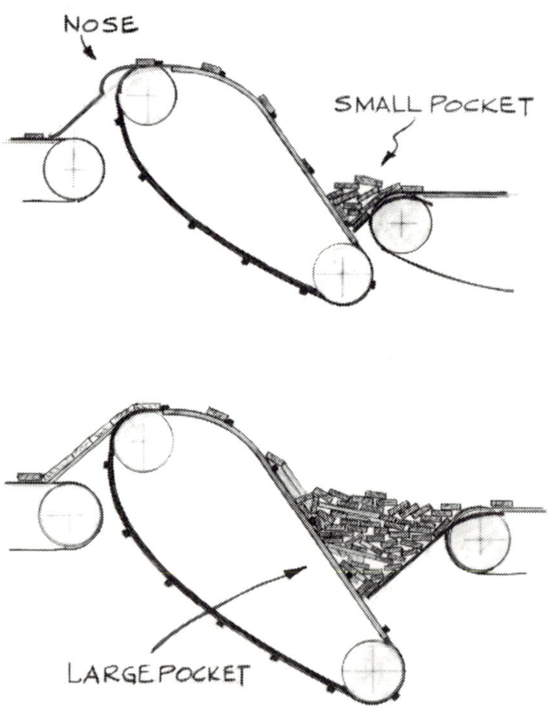

Figure 6.28

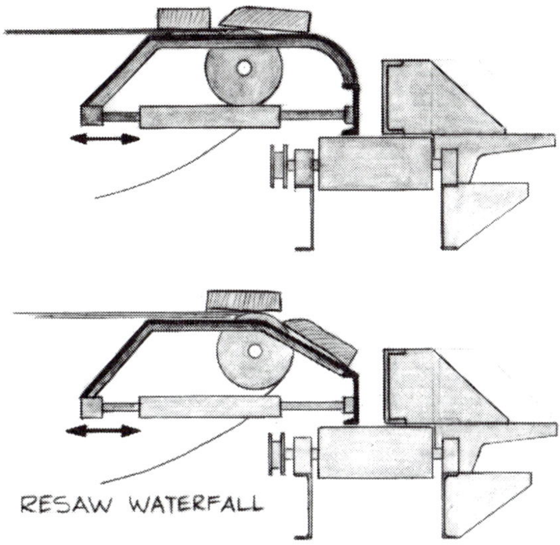

Figure 6.29

do what that name implies. This is not what it is doing, however, in the majority of the installations I have observed. There is a tendency to use this equipment as a surge bin. With the huge holding pocket, it now becomes a scrambler and a lumber breaker. Many thousands of feet of lumber have been destroyed in these pockets. In addition, maintenance on chains, bars, shafts, and bearings is high. Figure 6.28 illustrates two designs, with the one shown at the top the preferred.

One additional feature that we recently designed is a simple step nose plate at the top of the unscrambler to prevent the lugs from catching the lumber if it builds up on the skid plate. You may also consider using a variable-speed drive so the operator has a chance to cope with the flow rate. My main point is this: use this equipment to separate boards, not in an attempt to solve surge problems in the flow.

Resaw linebar. Figure 6.29 illustrates two different moving waterfalls. The top one, while nicely shaped, is designed incorrectly. The operator is compelled either to manhandle the lumber off the curved section or to retract the waterfall each time he wants to feed lumber down into the slot. This is not only slow, but it causes unnecessary work for the operator. The design at the bottom allows the operator to keep the linebar closed or stay 3 inches behind the sawline. The majority of his stock is such a size that it will merely slide down the waterfall and drop into the slot. If he wants to turn a piece, he does this just before it drops off the edge of the waterfall. To feed a wide piece, he operates with the waterfall retracted. I have shown this example to point out that there are times when we install incorrectly designed equipment and then train the operator to operate awkwardly.

Increasing conveying effectiveness

Much controversy goes on about conveyors. Here are some simple improvements:

Lap versus S-drives. Some like S-drives on long link chain; others like lap drives (Figure 6.30). My preference is lap drives. One complaint is that lap drives do not provide enough wrap. Use a large sprocket, a seven-tooth sprocket. Be sure it is lined up properly and kept in pitch. You will find with many S-drives that the chain is going back on the S and hitting the return chain. You are not getting too much lap in that case. The other problem with an

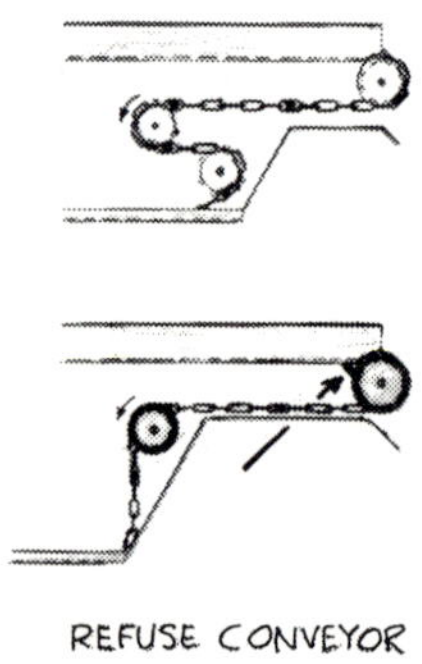

Figure 6.30

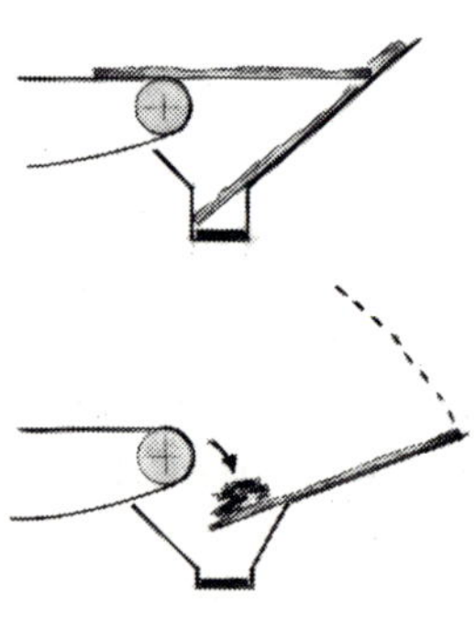

Figure 6.31

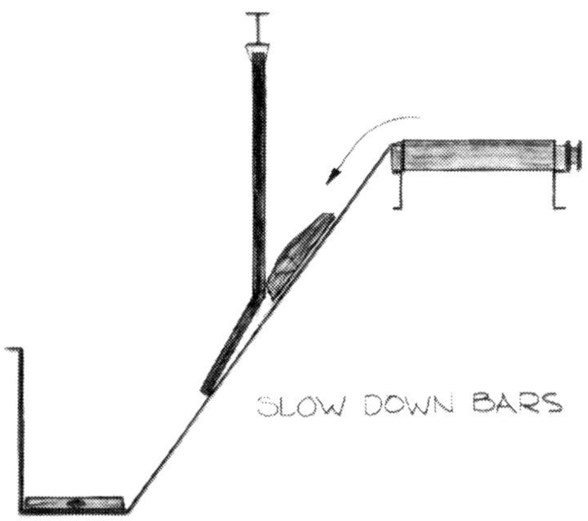

Figure 6.32

S-drive is that it carries the debris back behind the drum. In the lap drive, however, if you design your chute properly, the debris drops onto the chute and is pulled away with the chain to the tail end. The arrow in the figure points to the cleaner in the groove of the drum. This keeps the sawdust out of the groove of the drum so the chain comes back straight to the drive. It is not unusual in some mills *never* to see that chain in the groove.

Conveyor chutes. When long wood enters the conveyor chute, it butts up against the wall of the chute. It cannot go anywhere, so it pushes the steel out. We are suggesting (Figure 6.31) that the chute be dropped down to about 15 degrees. This permits long wood to be pushed out onto the flat surface with the chain. The debris coming from the conveyor will fall onto the end of the long piece and eventually will turn it around so it can proceed in the new direction.

Protecting conveyors. In many mills conveyors are pounded into almost unrecognizable shapes by heavy slabs being dropped off headrig roll cases directly to the basement. You can install a number of 1¼-inch by 4-inch bars (Figure 6.32) to make a magazine feed. It is simple, but it saves the conveyor. This is particularly necessary if you have vibrating conveyors.

Troughing belts. Most mills install troughing rolls in the loading areas. We do not feel this is a good practice. It is impossible to follow the belt with a skirt when the belt is sagging between the rolls. We suggest removing the rolls from the loading area completely and replacing them with formed steel (Figure 6.33). Be careful, however, that you do not form it so that the belt creates a suction. Put in a dropped bottom to create an air cushion, which must be maintained at all times. The air cushion takes the impact and also keeps the belt from being sucked down to the bottom of the trough. Now the skirt will fit because you have a straight side. If you want to put formed steel sides on the belt conveyor for its entire length, just carry it on. The portion by the top of the roll must be cut around the roll. These sides will prevent spillage when conveying cedar hog fuel.

Double pull belts versus Sweeds. Figure 6.34 shows a double belt, one above the other, to replace the undesirable electric Sweed. This is very effective. The angle of the belt and the speed of the belt are set in such a way that the travel forward and the

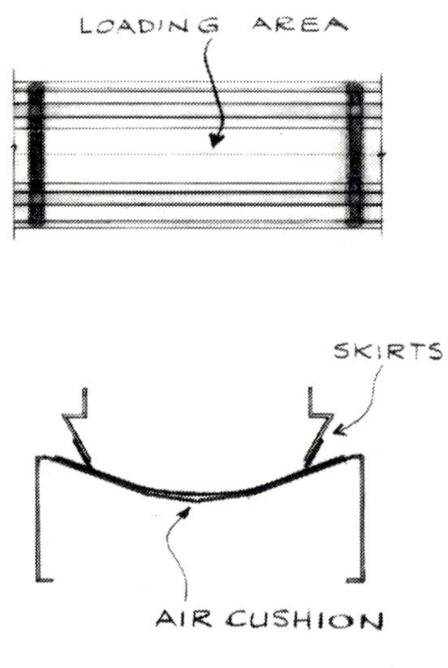

Figure 6.33

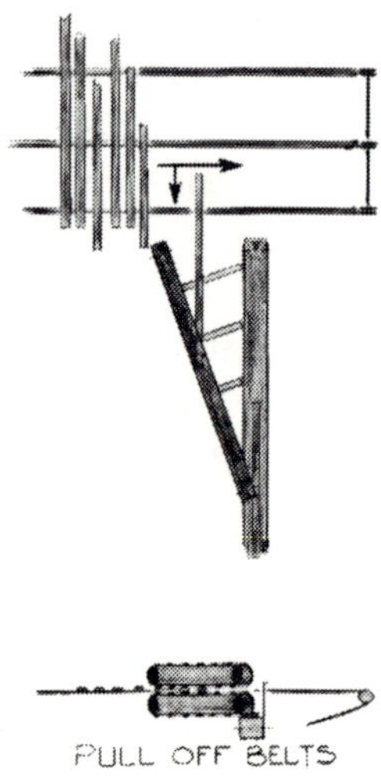

Figure 6.34

travel sideways match the speed of the transfer. This eliminates the "whipping" generally experienced with a Sweed. When an electric Sweed clamps down on a board, the transfer on which the board is riding continues to control the board until it gets halfway into the Sweed. The whipping that causes is very dangerous. The belt shown in the bottom sketch is fitted with cleats 2 inches high at 8-inch centers. These soft cleats are vulcanized onto the belt; they grab the lumber and pull it off the transfer with no problems. With these belts, one board can be sent off after the other, as tightly as desired.

Tilted-pan transfers. In many mills, conveyor after conveyor after conveyor of material is dumped onto transfers when they already are carrying lumber. Figure 6.35 shows an alternative—use of a tilted pan. Discharge lumber onto the pan and drop it off flat on top of the lumber already on the conveyor. Your reaction probably is that huge surges of lumber will tear it out. But I think if your mill is running correctly, you will reduce those surges of lumber. Incidentally, the picture is not quite correct because a board is extending out beyond the pan. The pan must be supported on the end somehow, so the boards will not be as long as the curl on the pan. Polyethylene can be used on the pan to make the lumber slide onto it with the greatest of ease. The pan also eliminates the need for straighteners on the transfers because the lumber will deliver itself very well.

Chromed and knurled rolls. Figure 6.36 shows use of a chromed, knurled roll—either a speedup, slowdown, or cant pushout roll—on the head end of a belt. Most roll cases and belt drums pushing lumber onto transfers are smooth. Many mills believe this to be a good thing but it really is not. How can you control lumber with a smooth roll? If you want lumber to go out at 300 feet per minute onto a transfer, you want to be able to control its speed. A chromed, knurled roll provides much better control. It should be chromed, naturally, because you want the knurl to last.

Chippers require exact setting

Chippers are precision cutting machines. If we want to produce high-quality chips with a minimum of fines and oversize, it is necessary to understand how to set up and maintain the chipper.

Figure 6.37 (left-hand circle) shows a chipper with the knife protruding too far (3/8 to 5/8 inch in some cases I have seen), and the knife clamps have somehow become set below the face plate 1/8 inch (right-hand circle). You can imagine the action in the chipper throat with the wood dropping off the heel of the knife, hitting the face plate, bouncing back and forth, and then finally

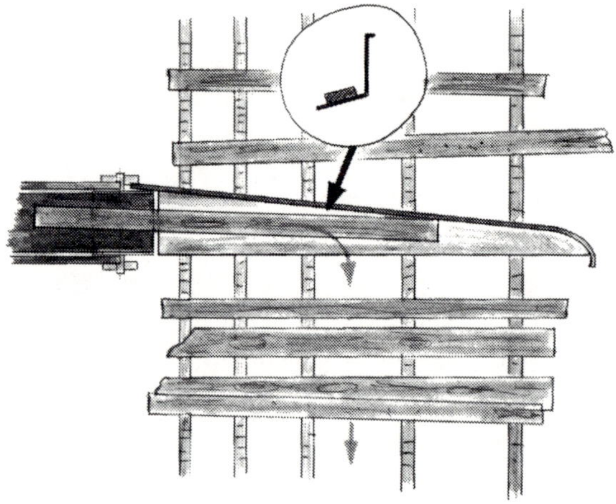

Figure 6.35

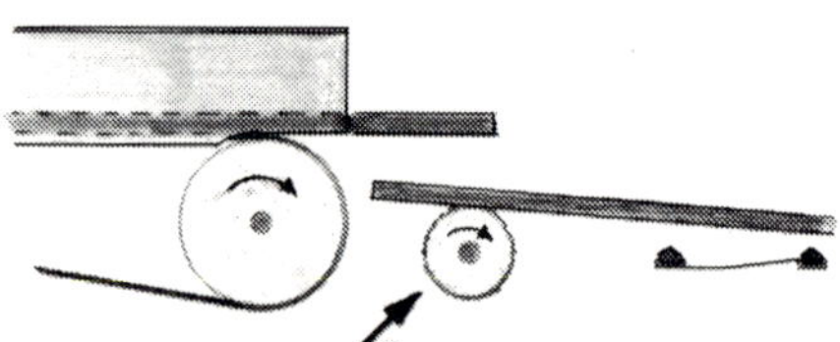

Figure 6.36

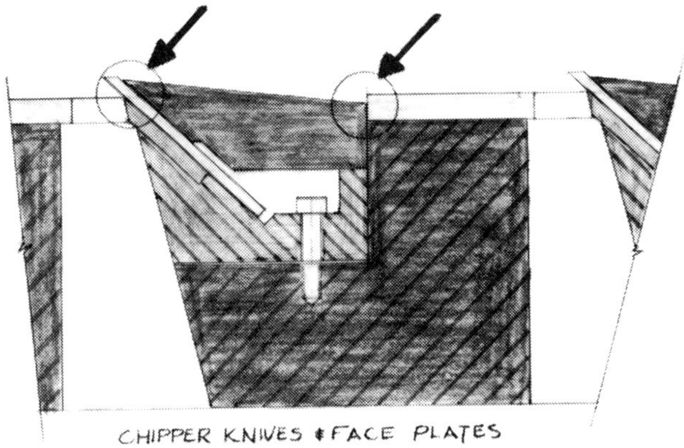

Figure 6.37

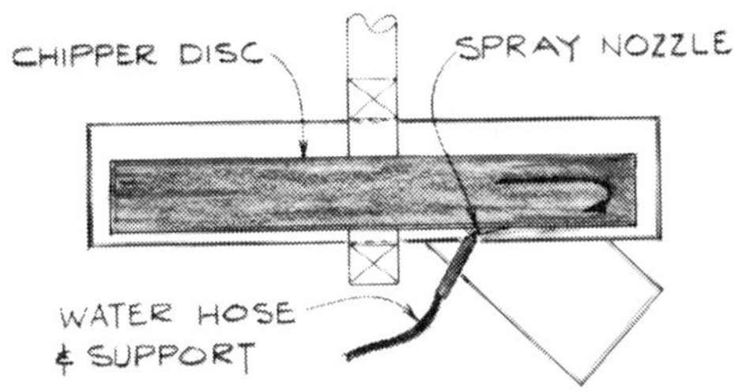

Figure 6.38

catching on the edge of the face plate and turning sideways. Two large chippers operating in one plant showed both these problems. The chip length was inconsistent and the oversize was excessive.

I might mention that in many cases chippers are purchased for 5/8-inch chips and later the knife is pushed out to cut 7/8-inch or 1-inch chips. To make this change, it is also necessary to purchase redesigned knife clamps that will fit within 1/32 inch of the heel of the knife. Many mills also forget that although they may be able to push the chipper knives out to cut 1-inch chips, those chips must go through the chipper slot. That slot is probably 3/4 inch or 13/16 inch, since it was designed to take a 5/8-inch chip. Trying to put a 1-inch chip through a 3/4-inch slot results in chips with curled ends—something the pulp mill does not like.

If there is bouncing going on in your chipper, you can hear it. If you do, you can be sure that you are losing feed capacity, as well as producing a poor chip.

Chipper face plates. A problem that often exists in the summer when chipping pine or fir is a buildup of pitch on the face plates. The roughness of the face plates will then flip the short ends up and make shingles. To combat this, simply put a small spray of water on the disc face (Figure 6.38). This also helps when chipping dry trim ends. Care must be taken to obtain the correct position and water flow rate.

Designing drop sorters for smooth flow

Nearly all mills of any size have one or more drop sorter stations to direct either finished lumber or lumber requiring remanufacturing. It is not unusual for these stations to curtail the production flow of the entire mill, particularly when the operator has four sorts. The lost time recorded in one mill for a year amounted to $20,401 for one drop sorter station. The type used in this mill was a series of three sloping arms which, when all closed, made up a long slide to the finished lumber transfer. This type, the cascade drop sorter, is second only to the flat live chain gates in slowness.

Over the years we have tried to improve the design of the drop sorter. The top sketch in Figure 6.39 illustrates a conventional drop sorter used in many mills. The problem with that type of sorter is that the main arms lap the head sprocket. This forces the operator to stop the lumber back far enough to allow the arms to swing open without hooking the lumber. If you watch the operator, you will find that he manhandles the lumber over the head sprocket.

The bottom sketch in Figure 6.39 shows a curved plate fitted over the headshaft of the transfer and extending forward to form a short slide. The main gate of the drop sorter now is set so it is below this curved nose. The top of the drop sorter arm is shaped

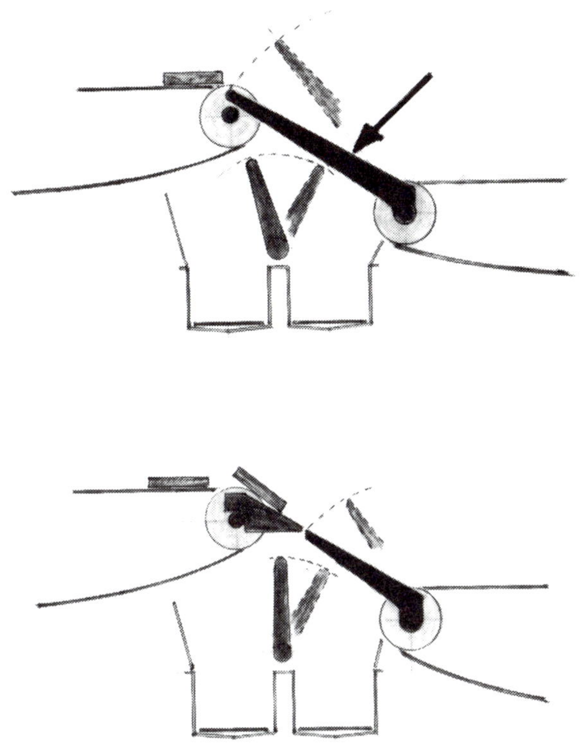

Figure 6.39

in such a way that it will stop the lumber as soon as it protrudes above this nose. This design has helped the operator considerably, as now he can use the headshaft to push the lumber out without having to manhandle it. You will note we show all our belt troughs with a V bottom. We developed this idea some years ago to prevent damaging the belt when dropping heavy lumber onto it. With a 16-inch trough, use about a 1½-inch drop; for 2 feet, a 2-inch drop, and so forth. Figure 6.40 illustrates a sorter design we have used on our last five installations. As shown in the figure, the design can permit three or four sorts. In this design we have

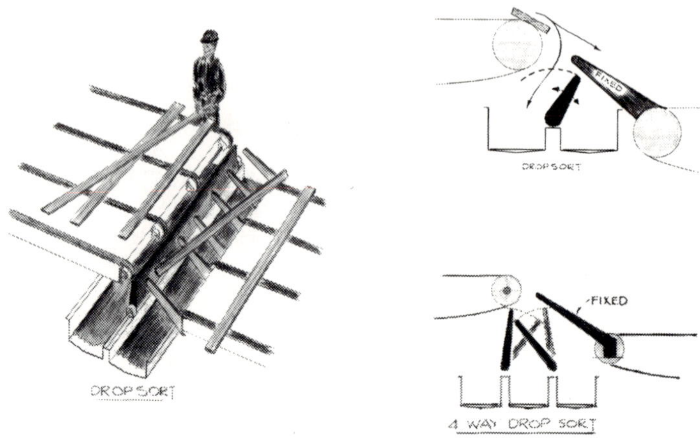

Figure 6.40

stationary main arms extending up to approximately 6 inches from the head shaft. The operator's function now is this:

1. For lumber he wishes to pass over the transfer, he merely holds back the end next to him. The angled board transfers over the gap because of the board being skewed.
2. To drop onto the belts, the lumber is straightened and then the transfer conveys it over the head shaft into the gap. By making the main sort arms stationary, maintenance has been greatly reduced. Some operators prefer polyethylene strips on the top of the arms to reduce the slope.

Another important feature is that this design has eliminated one foot button for the operator. We consider this sorter to be 15 to 20 percent faster than even the improved unit we showed in Figure 6.39. Of course, if you wish to sort timbers, it will be necessary to use the sorter with the main gate opening.

7

Control logic functions inexpensively and precisely

Robert P. Lewis, Engineering Manager, Lewis Electronics, Inc., Hillsboro, Oregon

Most of you are aware of the revolution going on in the sawmill industry as far as sophisticated setworks, sophisticated scanning systems, sorters, and things of that nature are concerned. I would like to call your attention to the small, menial tasks that are taking place in your mill. As recent newspaper ads indicate, the price of computers is falling drastically. You can now buy a small hand-held computer that will perform a variety of mathematical tricks and stunts for $30. Many of the advances that have taken place in the electronics industry in development of sophisticated hand-held calculators are now finding their way into equivalent types of small processors and computers for industrial control applications.

The point is that recent developments in integrated circuits are making very powerful logic systems available to everyone at bargain-basement prices. The word *logic* is the key. Control units and systems used throughout the sawmill are embryonic logic systems. For example, a limit switch wired to activate an air valve solenoid is a logic function. The wiring that has gone into the switch and the interconnections between the various relays are the "programming" of that little "computer." The calculators now available do their job by taking a "universal" logic element—the CPU or central processor unit—and running a sequence of instructions—the program—through it. Different parts of the program respond to the pushing of different keys.

Developments in the hand calculator have also had a great impact on larger computers, bringing them through development phases from the large-scale computer to the minicomputer to the microprocessor. We are now applying fourth- and fifth-generation "computer-on-a-chip" units to many jobs in many areas.

Determining the power of a logic element

The universal logic element is now a very real and useful tool for many jobs. These include control, as well as what the computer industry calls "number crunching." The power of the logic element is a function of basic factors in its design:

1. *Instruction repertoire:* The number of different types of instructions the logic element is capable of taking is the instruction repertoire.
2. *Word size:* Many microprocessors now on the market are 8-bit word size. Some are 16, and we have seen a few advertised with as high as a 32-bit word size.
3. *Cycle speed:* The number of different types of instructions the logic element will follow times the number of instructions per second it can execute determines how fast it will complete a given program. Perhaps you have a certain number of things you want the logic element to do. You want it to do each one in sequence and get back to the first task before, for example, something has run off the end of a conveyor. The cycle speed of the machine—its ability to get back around and to keep track of that board that is headed off the conveyor—is an important measure of its power.
4. *Address capability:* The fourth item that defines the power of a computing system is its address capability—how big a program can be put into the computer. This has to do with the number of instruction sets that can be stored in its memory. It must be able

to go through these addresses, fetch the instructions out of memory, execute them, and then step to the next instruction.

The computer we might apply to a particular job will consist of several parts, the most important of which are the *black box* or *universal logic element* and the instructions (*program*) it is to execute.

Black box (hardware) = The "body" of the control system
Instructions (program) = The "personality"

It may seem like sending a sledge hammer to swat a fly to use a computer to replace a few relays, but that is exactly what I am advocating. One advantage of this approach is that if the system has more logic power than you need, it can be expanded very easily to take on other tasks as you later define additional control functions you want it to perform.

The universal black box can simplify sparing, where a single microcomputer plug-in card can plug into any one of many systems. After it gets into the system, the program that is stored there gives it its personality and tells it the task it is to do.

For example, we are running the identical computer in four different applications. In one case, it is a lumber sorter; in another, it is a setworks on a carriage; in the third and fourth cases, it is a setworks on a linebar resaw and on a four-saw edger. Basically, the only difference among these systems is the program we have plugged into the computer. The computer's physical hardware and the cards that are plugged into the card cage are fundamentally identical from one system to the other. This particular characteristic is the reason computers deserve to be looked at for simpler tasks than those for which they have been considered in the past. People sometimes get carried away with the computing capability they have. They are looking always at bigger and bigger systems and, I think, letting some small things sneak through the cracks.

If you have the universal hardware—the black box and the various cards that go with it—the same basic computer with one set of sparing cards could serve to spare units in several mill areas.

Hardwired control versus computer control

To give you an idea of the simplicity of computer application, I will compare how an extremely simple task might be wired with how it might be controlled with a computer.

In Figure 7.1, we have a very simple task. We want to set two limit switches and cause a relay or a valve solenoid or something similar to respond when both limit switches, the A switch and the B switch, are closed. In computer language, this is called an AND circuit. A *and* B must be closed for C to happen. That could be directly hardwired in your mill by stringing a wire from one switch to the next and around and over to the relay it is to operate (No. 1 in Figure 7.1).

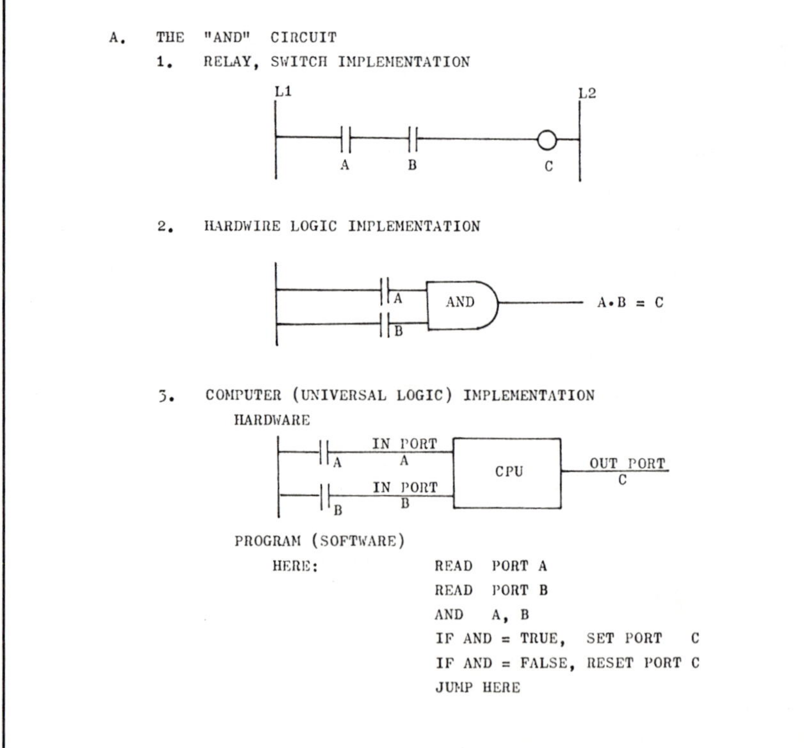

Figure 7.1. Three methods of controlling an AND circuit operation.

A second method would be to install dedicated logic, a particular type of integrated circuit called the AND circuit or AND gate, and wire the two switches into the AND gate. The AND gate would perform the logic functions of detecting that both A and B are closed and would output a signal to C.

In the computer example, we have A and B feeding into the computer, just as we did in the hardwired system, but we construct a program to tell the computer to read the A switch, then read the B switch, then perform a logical AND on those two signals. The computer says, "Both A switch and B switch are closed." The next step in the program says that if the AND is true (if it is occurring), set the output C. If the AND is false, reset the output C and we have accomplished the same thing. After the computer has completed the steps outlined, it jumps back to the beginning of the program and continues to run in a circle, constantly sampling the A and B switches and constantly outputting a signal that is the result of the logical combination of the two switches.

In Figure 7.2 the two switches have been rewired slightly so the two switch contacts are now wired in parallel. This is an OR circuit. If switch A *or* switch B is closed, then set relay C. In the hardwired examples, some type of integrated circuit that performs the OR logic function is used. But notice that in No. 3, the wiring on the computer is identical with that in Figure 7.1. A switch and B switch wire into the computer just as they did in Figure 7.1 and the C output is wired to the same place. The only difference between the two functions, the AND function and the OR function, is that the program tells the computer to respond if A *or* B is closed and to continue repeating the loop and sampling it. These are two simple applications of the computer to perform a logic function.

Controlling a veneer stacker

The circuit at the top in Figure 7.3 is a wiring ladder diagram that would be used in a veneer stacker for automatic activation of the drop—when a piece of veneer comes into a bin and is to be dropped onto a stack. The usual situation is to have the front edge of the veneer strike two limit switches, one at each edge of the stacker, that disengage drive clutches. The clutches are used to straighten the veneer. When the leading edge of the veneer strikes the first switch, that clutch is disengaged. When the leading edge of the opposite side of the veneer strikes the other switch, the clutch on that side is disengaged. Then the drop cycle is activated and the veneer dropped onto the stack. A drop timer controls the cycling of the drop arms.

To perform the same function with a computer, the wiring becomes simpler. In the top example in Figure 7.3, we had a total of four contacts plus a timing relay with a set of contacts to drive

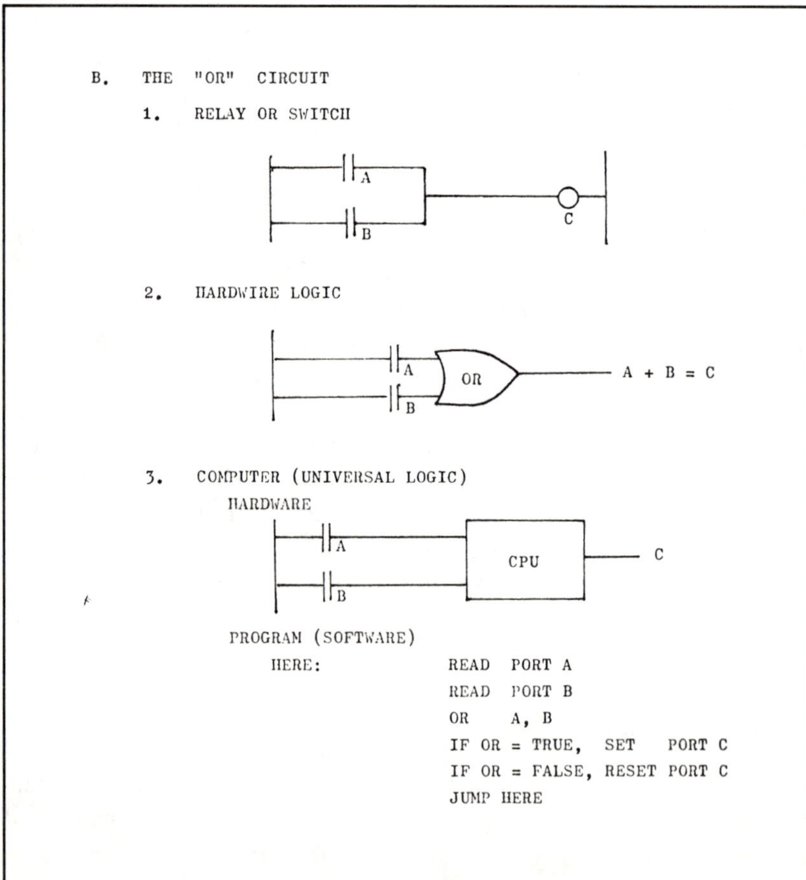

Figure 7.2. Three methods of controlling an OR circuit operation.

APPLICATION EXAMPLE: STRAIGHTENING STOCK IN A STACKER

(2) INPUTS LEFT, RIGHT LIMIT SWITCHES

(3) OUTPUTS LEFT, RIGHT CLUTCH, DROP SOLENOID

1. RELAY or SWITCH

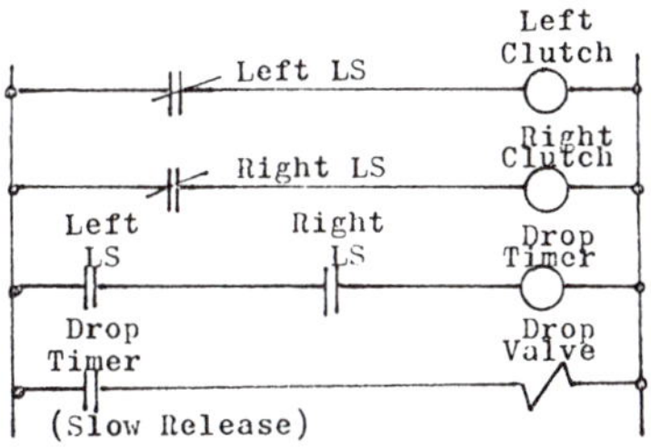

2. COMPUTER/UNIVERSAL LOGIC

HARDWARE

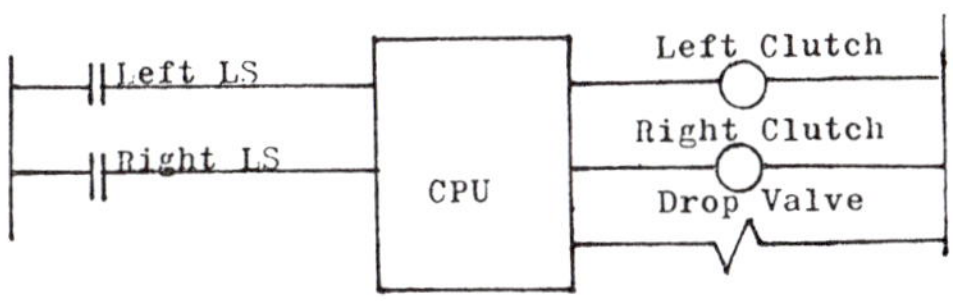

PROGRAM (SOFTWARE)

```
START:      READ LEFT LIMIT SWITCH
            IF UNOPERATED SET LEFT CLUTCH
            ELSE RESET LEFT CLUTCH
            READ RIGHT LIMIT SWITCH
            IF UNOPERATED SET RIGHT CLUTCH
            ELSE RESET RIGHT CLUTCH
            AND LEFT, RIGHT LIMIT SWITCHES
            IF AND = TRUE, SET DROP VALVE
                & JUMP TO TIME
            ELSE RESET DROP VALVE
            JUMP START

TIME:       DO A WAIT LOOP FOR TIME
            RESET DROP VALVE
            JUMP START
```

Figure 7.3. Hardwiring versus computer control for straightening stock in a veneer stacker. Computer program is based on AND circuit.

a drop valve. We also had to drive the clutch valves to disengage the clutches. With computer control, we wire two switches into the computer and we have three outputs from the computer. These outputs drive the left clutch, the right clutch, and the drop solenoid or valve. No timer is involved.

The programming simply tells the computer to read the left limit switch to see whether it has been activated. If the left limit switch is not operated, meaning it has not been struck, keep the left clutch activated and driving the conveyor or belt system. If the left switch has been activated, reset the left clutch, or disengage it, to stop that side. The computer would then read the right limit switch, going through the same process. When it finds that both left and right limit switches are set—if the AND of those two switches is true—set the drop valve and jump into a subroutine called "time." If the AND is not true, reset the drop valve, jump back to start, and keep scanning the switches.

The "time" subroutine is this: We install a number in a register of the computer. We have the computer decrement, or count down from, that number. When it reaches zero, the task is finished. It then resets everything. It has timed itself out and is ready to accept the next piece of veneer.

Simple programs like these are extremely easy to execute with the level of computers on the market today. These computers can perform the steps so rapidly that we could have a computer doing the identical function for several bins simultaneously. It does it fast enough that the other bins would not know the time was being shared among them. A program of the type shown in Figure 7.3 could be executed by the computer (excluding the timing out) in a matter of 20 millionths (0.000020) of a second, giving it ample time to look at other bins and do other jobs.

My point is this: The computer can be a universal control element. *Computer* is merely a new word. You have been performing logic functions with the control systems now in your mills—relays and other types of electronic devices. Since the computer has flexibility, it can be sent to do many jobs. If it can be sent to do many jobs, then it can greatly simplify sparing on operating hardware you have. You can combine several tasks on a single computer because it is fast enough to service each task and be back around by the time each task next needs its attention.

Interrelationship of system elements

The block diagram in Figure 7.4 shows how a computer system fits together. The computer consists, functionally, of the CPU (central processor unit) and of a series of cards or boards that plug into it. These perform what we call *interfacing functions.* The interface cards adapt the signals from the outside world to the signal format with which the computer works. Or in the case of output cards, they adapt the computer format back to what is required in the outside world.

Sensors can be placed in the outer world. These can be switches; camera systems; scanners; or encoder devices (which keep track of the rotation of shafts or the motion of chains). If you want to enable a human operator to talk to the computer and tell it certain things, the system can be equipped with a keyboard. Many types of keyboards are available. In some cases, a keyboard looks like nothing more than a conventional operator control station. In a more sophisticated installation, you might have a teletype, a high-speed printer with a keyboard, or even a tape reader. The input interface then takes signals from these sensing devices and converts them to the format the computer likes to work with, the language it understands.

The central block is the central processor unit (CPU) plus the program, or set of instructions, prepared to make the CPU do the tasks assigned to it. If necessary (and many times on smaller tasks this is not necessary), a type of memory (a random-access or read-write memory) can be used to accumulate information about what has been happening while the task was being accomplished.

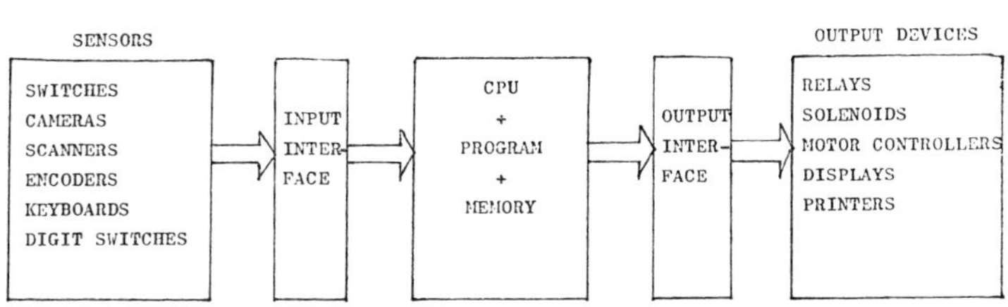

Figure 7.4. Interrelationship of computer system components.

To accumulate board tallies or log tally information, you equip the computer with sufficient memory to store those numbers and recall them later in a printout, or report.

Output devices run the gamut of control devices in use in the electrical industry today—relays, solenoids, motor controllers, numeric displays, and printers. I cannot really set a limit on the number of output devices one could put on a computer, so long as you have prepared a program that tells the computer what it is supposed to do with each device—how it is supposed to service it. Each device tied to the computer must have a corresponding service routine or program stored in the computer to make it perform its assigned task.

Functions of the interface devices

The voltages with which the computer likes to work and the signal format it likes to see are not necessarily what the outside world wants to give it. Most microprocessor systems in operation today are very low voltage units, as low as 5 volts. The systems also have noise immunity problems—electrical interference noise can talk to them. The job of the interface devices is to convert the 5-volt signals from the computer to higher voltages to drive the valve solenoid or to drop the higher voltages from a switch to a 5-volt signal so it can talk to the computer. Interface devices also perform a function called *signal conditioning*, which is to screen out noise so the computer does not have to listen to the arcs on a relay across the room or a chattering solenoid with a loose wire. Proper interfacing and proper signal conditioning are among the most important aspects of tying a computer into an industrial control system.

Our computers have a series of interface cards with the various signal conditioning aspects designed into them. One frequent problem involved limit switches. The contacts are not necessarily closed simply because the lever has been thrown. Contacts bounce for a short time. One problem computers have is they are really too fast. The switch closes, so the computer responds and looks at the switch. The switch is no longer closed, because the contacts just momentarily bounced open. That signal can be treated to remove the switch bounce and chatter. The bounce occurs for only a few thousandths of a second. One solution is this: When the switch first closes, have it tell the computer, "I am closed." The computer responds, "All right, I am not going to pay any attention to you for 0.015 second." After that time has elapsed, the computer rechecks the switch. "Sure enough, switch, you are closed. You must have meant what you said. All right, I will do whatever is to be done when you are closed."

Things of this type are built into the signal conditioning aspects of the computer. They can be further heightened by techniques in programming. You may have the computer look at the switch several times to be sure each time it is checked, it is still closed. That way, the computer can make a logical decision that the switch really meant what it said and respond to it.

You might think that card upon card upon card must be plugged into the computer to tie it to the outside world. That is really not so. The interface cards for essentially all microprocessors on the market today are up to 16 bits, or signals, wide. That means that one card can respond to 16 different pushbuttons or limit switches. An output card, likewise, can drive up to 16 different outputs (Figure 7.5).

Computer hardware described

Hardware currently used to accomplish various tasks includes the keyboard, which allows the operator to instruct the computer which parts of its program he wants executed. If you look inside a central processor unit, you will see the physical hardware that responds to the instructions given by the operator through the keyboard. A test panel with built-in metering and status display lamps tells which inputs and outputs have been turned on or off. Immediately below the status panel is a space where numeric readouts can be installed. These give readings of the functions taking place in the computer.

Whenever a control system or a computer is put into any environment, provision must be made for changes. Mill operating personnel may decide they want a different type of measurement on a board, for example. Wheel switches are often used to permit changes in variables. The dimension parameters with which you want the system to work can be dialed in readily. Providing five

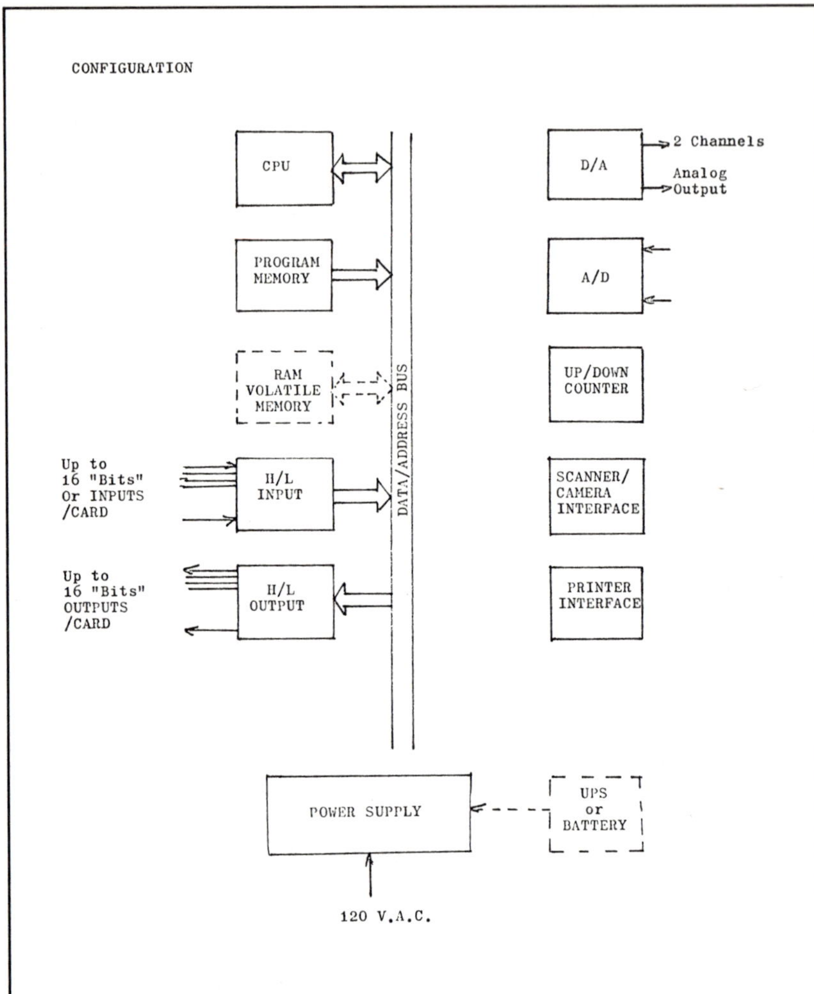

Figure 7.5. Typical computer system configuration.

digits allows you to go from 99 inches to zero in thousandths-of-an-inch increments. Other thumb wheels are utilized to establish timing sequences. You dial the time with the thumb wheel. The computer takes that number and performs the countdown program described earlier.

The physical computer is one single chip on one of the cards. The rest of the cards inside the housing are interface cards that tie the computer to the physical outside world, let it sense what is happening, and let it perform its functions. A few cards will contain the program memory, the set of instructions that have been plugged into the computer to give it its personality.

We try to build a large amount of what we call BITE (built-in test equipment) into computer systems to facilitate diagnosing trouble. We like to think our product is perfect, but this is not always the case. The more we facilitate troubleshooting in the mill, the more we reduce downtime. If the BITE has been designed properly, the engineer or the serviceman may not even need to come to the mill. He can say to the mill's operator, "Push such-and-such a button. Did light so-and-so go on? If it did not, replace card thus-and-such, at the bottom left."

I hope I have given you a little more appreciation of what the computer is. It is no longer a magic, space-age device from never-never land. It can be a very useful working tool for you.

8

Raise sideboard recovery with computerized edger

Chris Vadnais, Sales Manager, Saab-Totem, Inc., Seattle, Washington

The Saab Edger Optimizer was first presented to the Northwest forest industry at the sawmill seminar held by Totem Equipment in November 1974. This was only one of a number of new sawmill products introduced by Saab-tement of Sweden and Sateko of Finland. Due in part to the enthusiasm shown for the edger optimizer at the seminar and to the desire to manufacture, market, and service this product through a Northwest company, Saab-Totem, Inc. was formed in 1975 as a subsidiary of Saab-Scania and Totem Equipment Co. of Seattle, Washington.

Saab-Scania is a major Scandinavian company with sales of more than $1.3 billion. The company is divided into five divisions: (1) aerospace; (2) cars; (3) Scania trucks; (4) industrial valves and instruments; and (5) Datasaab computer division.

Board edging in the sawmill has seen very few improvements. In 1973 the Datasaab division started to develop an automatic and computerized edging system. To date, Saab-Scania has spent more than $2 million in research and development for the optimizer.

The market is changing. Companies desire to buy complete systems, including both mechanical and computer functions, with a fixed cost and guaranteed performance. By computerized scanning we guarantee 92 percent of the maximum sideboard volume

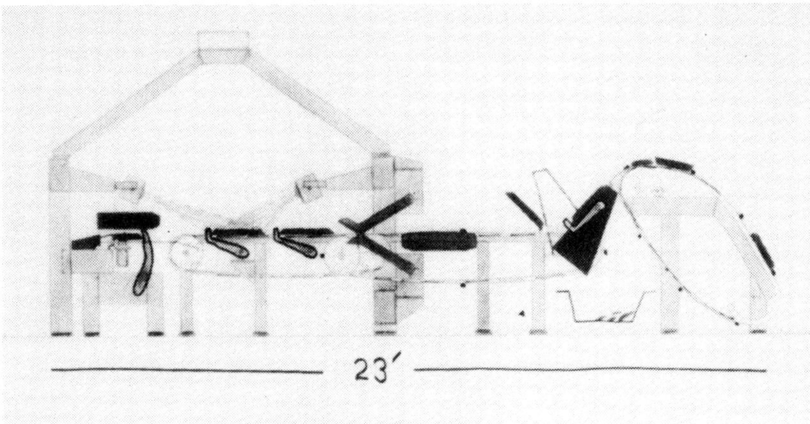

Figure 8.1. Diagram of the edger optimizer, designed to handle 15 boards per minute. Main components of system are shaded. From right to left: board feeder; endlining device; scissor-like thickness measurer; scanning zone; servo cylinders for skewing and centering.

and 15 boards per minute through the edger. A diagram of the prototype edger optimizer built for the North American market is shown in Figure 8.1.

Edging can be divided into four parts:

1. Feeder, with reject and overturn function.
2. Scissor-like thickness device.
3. Scanning zone.
4. Skewing and centering of the sideboard by servo cylinders.

The edger optimizer is designed to take sideboards from a quad or Chip-N-Saw. The edger optimizer must be fed from one side only or singulation and scanning must be duplicated on both sides for a similar rate. It can be run in a computer mode or an independent manual mode.

Few decisions required of edger operator

First in the sequence of operation, the board is transferred from the buffer storage zone up the unscrambler, which is powered hydraulically and controlled by photocells. The unscrambler angle is at 60 degrees, with individual lugs spaced 2 feet apart. The unscrambler has a creeper speed to ensure that if two boards are brought up, only one is placed in the feeder.

The board is held in the feeder for the operator's decision: (1) to proceed; (2) to overturn, so the wane is up; or (3) to reject. The operator also tells the computer the highest quality of the board. These are the only decisions made by the operator in the computer mode.

The board then is placed on a lug chain, with final endlining, and is transferred through the 10 pairs of scissor-like thickness measuring devices, which have a resolution of 0.010 inch. These are spaced on 2-foot intervals and used to check the setting accuracy of your quad or Chip-N-Saw, as a quality control. The thickness devices will also be used in the future edging of slabs for either a top chiphead or a resaw.

Next, we have a waiting position before the scanning area.

How boards are measured

When the sideboard passed through the thickness measuring zone, we measured its basic length to know which *two* of the *four* fixed stops in the scanning zone to raise. These line up with the two corresponding servo cylinders used to skew and center the sideboard on the infeed chain.

The board is illuminated from one side in such a way that the edge closest to the light is detected when the scanning array gets light reflected back to it (Figure 8.2). The top edge on the far side is also detected when the board is lit from this side, with the scanner detecting the change in light intensity from the top of the board (which is very bright) to the far sloping side or wane. This slanted surface does not reflect light back to the scanner because it is cast in a shadow. This process is then reversed when the light source is switched to the other side. We now have all four points on a cross-section.

Datasaab is using two 512-bit Redicon arrays, each having a scanning envelope 20 inches wide and 12 feet long; thus it can scan a 24-foot-long board. The four points gained on the cross-section every 4 inches have a resolution of plus or minus 1 millimeter each.

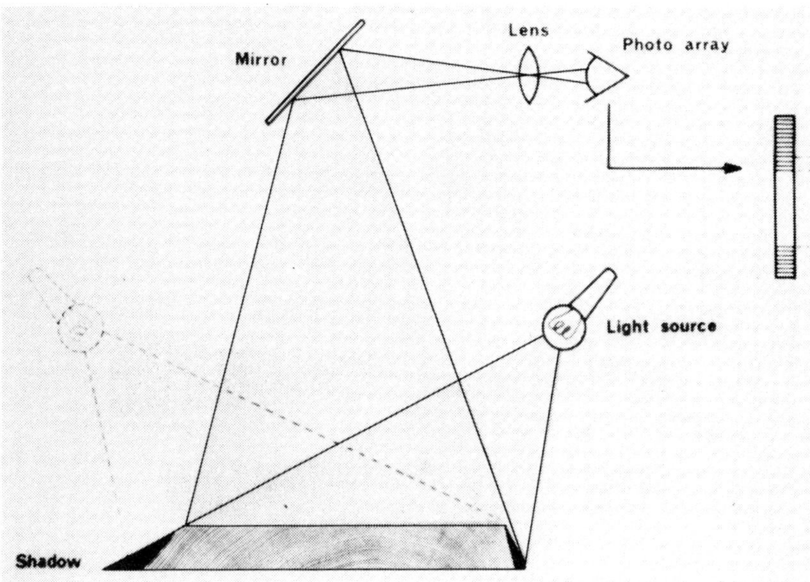

Figure 8.2. Edger optimizer scanning system uses reflected-light principle to measure boards. Boards can be up to 24 feet long.

The board is scanned every 4 inches full length by a mirror on a stepping motor. A diagram of the measuring operation is shown in Figure 8.3. With this information, the computer finds a reference line for it and the edger to work from, on either a full taper or a half-taper basis.

Computer evaluates input and specifies edging

In the next step, several board width alternatives and combinations with the splitter saw are tried. Maximum volume is calculated for each, from the lowest quality to the highest quality given by the operator at the reject feeder.

Input to the computer for evaluating the volume and price relationship is shown in Figure 8.4. A number of different inputs are fed into the computer: (1) the edging rules, which are fixed; (2) the market situation, which you are changing, say, weekly; (3) the thickness measurement, so the computer knows whether it is looking at a 1-inch, 2-inch, or 4-inch board; (4) the information coming from the actual scanning head itself; and (5) the quality

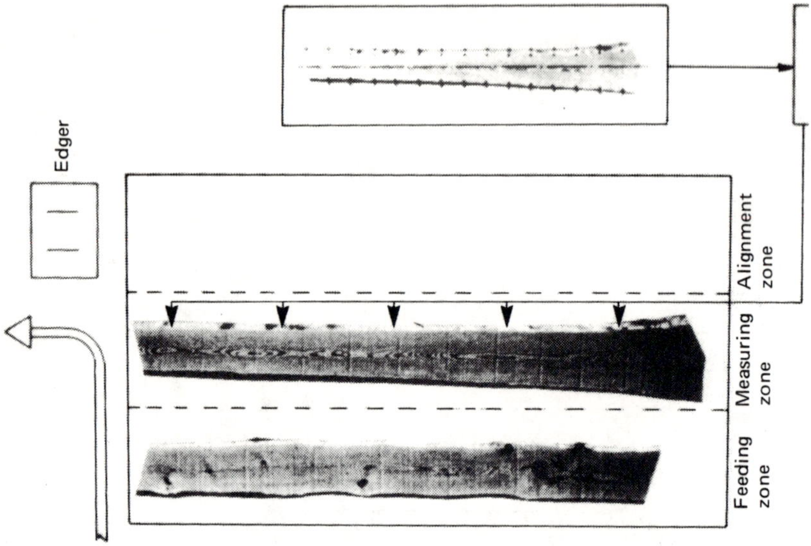

Figure 8.3. Scanners measure board at 4-inch intervals with a resolution of plus or minus 1 millimeter full length before it enters the alignment zone in front of the edger.

instructions given by the operator, so the computer knows the maximum quality it can look at. It starts at the lowest quality and then proceeds consecutively to the highest quality given by the operator.

The dots on the board in Figure 8.5 are the scanning points. The resolution of those dots is plus or minus 1 millimeter. The computer is trying several least squares fit lines (shaded in the illustration). It is looking at the maximum volume for that particular quality and at the price it gets for that particular quality with reference to the volume contained in it. What the computer is trying to come out with is a maximum dollar value for this particular board. Here the computer has selected a higher volume and a lower price to achieve maximum dollar value. It could have chosen a slightly lower volume and, consequently, a somewhat higher price. But this would not yield as high a dollar return. These are the alternatives at which it is looking.

A specific board with a production factor of 100 percent is judged solely on its highest market price; but with a production

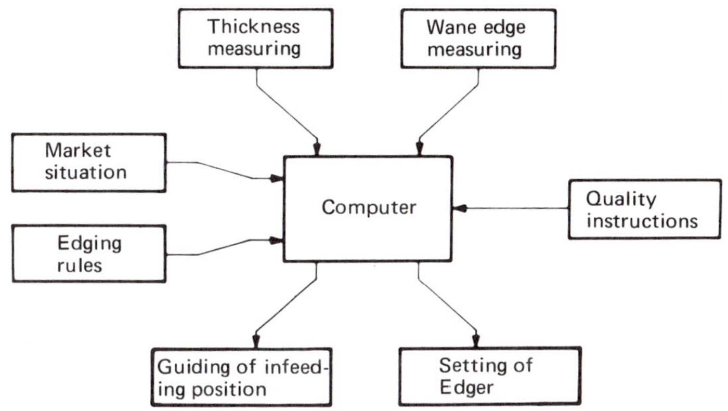

Figure 8.4. Computer considers a number of factors in evaluating each board for the maximum dollar value.

factor of 120 percent, the computer will be influenced to choose it over another board which would have had a slightly higher price. For example, if the market on a particular board is flooded and you decided you do not want to generate it in your mill, you can put a lower production factor on that board—say, 70 or 80 percent. If an item is in demand and you must have it to fill certain orders, you would put a higher production factor on it.

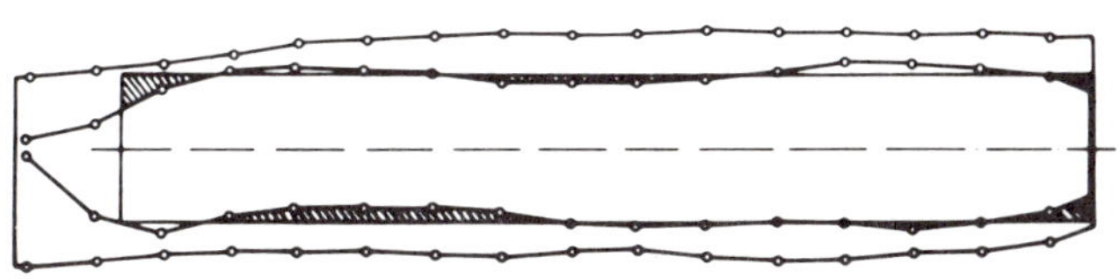

Figure 8.5. Using scanning data, the computer fits various edging alternatives to board under consideration. It may choose, as here, to maximize dollar return rather than wood volume.

Chain replaces roll case at edger infeed

With the scanning completed and the saw or chiphead sets given to the edger, the board is transferred to the infeed chain of the edger where the servo cylinders have already been set by the computer for skewing and centering the board.

A single chain infeed eliminates the fiber creep that occurs when many transfer points, such as a roll case, are used. Elimination of the roll case also solves the problem of keeping a roll case in alignment.

With the board on the infeed chain, a recessed arm presses the board against the servos, the hold-down rollers are lowered onto the board, the recessed arm is retracted, the 2½-inch-wide 18-foot-long infeed chain is raised, and the board proceeds into the edger. The transport through the edger was also redesigned using a 6-inch-wide quad-80 chain for both the infeed and outfeed sections of the edger.

Since the board is centered on the chain, it is necessary to move both chipheads or saws with respect to the centerline of the infeed chain. In the manual mode, the board will be centered by four sets of calipers on the infeed chain, with the operator choosing the width and offset function. The offset is used if one side of the board has 2 inches of wane and the other only 1 inch. The feeling here was to have the manual mode as simple and uncomplicated as possible—customers do not spend this kind of money for a computer system and expect to run it manually.

Replacing chips with boards offers sizable dollar gain

This brings us down to the bottom line—what is the return on your investment?

What we have been talking about is the conversion of material presently going to chips into boards. Studies conducted by the Swedish Research Institute found that the "best" manual operator was getting 85 percent of maximum volume; the worst, about 70 percent. Let us take the high of 85 percent. We are increasing this only 7 percent—from 85 to 92 percent. Approximately one-third of a mill's production is sideboards. (Again, this is from a quad or Chip-N-Saw.) A small-log mill producing 60 million board feet would have 20 million board feet coming from sideboards. Twenty million board feet increased 7 percent is an extra 1.4 million board feed that would have gone to chips.

If that 1.4 million board feet goes to chips, the return is approximately \$20 per thousand board feet (when 1,000 BF = 1/2 bone-dry ton of chips). If, on the other hand, that 1.4 million board feet becomes boards, it is worth around \$150 average per thousand.

20 MMBF increased 7% = 1.4 MMBF
\$150 boards – \$20 chips = \$130
\$130 X 1.4 MMBF = \$182,000

The difference here of \$130 is almost all profit. The mill is running anyway; the costs are incurred whether you get this extra recovery or not. The only additional costs would be from the increased volume at the dry kiln, planing, and packaging departments. This increase in recovery (ranging from a minimum of 7 percent) comes from getting that 4-inch-wide board you are not presently able to get, and going to the next wider width and still meeting the grade rules. To do this requires extreme accuracy.

The first edger optimizer has been sold to a major Northwest company. It has been designed and manufactured by Saab-Totem, Inc. of Seattle, who is responsible for all aspects of its performance and operation.

9

Points in selecting a wastewood chipper

Fred Veuger, Vice President-Marketing,
CAE Machinery Ltd., Vancouver, British Columbia

Chipping of wastewood is distinct from log and headrig chipping. Let us discuss the selection of a wastewood chipper by type and size. Then I will cover some of the machine features. Proper selection of the chipper may well make the difference between a profitable or an unprofitable situation.

Many chipper problems in mills can be attributed to lack of attention to details when the machine was selected. It is natural for the mill manager and his engineer to spend a great deal of time selecting the type of sawing process and the machinery to accomplish this. Yet chip revenue plays an important role, particularly during poor lumber markets. Modern mills are designed with chippers close-coupled to the sawing process. Any problem at the chipper will shut down the headrig within a few minutes.

Base chipper selection on wastewood handled

Many of these problems can be prevented. The first step is to analyze carefully the type of wastewood which will be chipped. Quite often, mill waste will be a mixture of various types of wastewood. A separate chipper to suit each type would be ideal, if economically justified. If not, a compromise should be made.

Long material, such as slabs and edgings, can best be processed in a horizontal-feed chipper. Some mills slash the long wastewood into shorter pieces. This tends to increase handling problems and will decrease chip quality substantially, as more ends are chipped.

Short, relatively heavy blocks, such as from a shingle mill, can best be processed in a gravity-feed machine. While gravity-feed chippers used to be standard in the industry, they have several disadvantages. These include the great amount of headroom required and, generally, a need to cut the wood into shorter lengths. Transition chutes to the spout must be designed carefully to avoid jamming of wood in the spout.

Ribs welded into a vibrating conveyor will help align short trim ends, improving chip quality. Tilting the conveyor also helps.

Perhaps the most difficult material to chip is short trim ends. These have a bad habit of jamming at the transition between the infeed conveyor and the chipper spout. Very short trim ends will tend to lie in the dead spot of the chipper spout. Trim ends will rarely enter the chipper with the grain parallel to and against the anvil. Unless they do, chip quality will deteriorate.

The introduction of vibrating conveyors and their use with chippers has eliminated or lessened these problems considerably. Unlike a belt conveyor, the end of a vibrating conveyor can be designed to be within a few inches of the chipper knives, thus avoiding a dead spot.

By slightly tilting the vibrating conveyor, positioning and alignment of the trim ends may be improved considerably. The chipper knives now have a chance to cut the wood against the anvil and at the correct angle, thus reducing overs and fines. Another way of aligning shorts is to weld ribs or angles into the conveyor. The latter method, however, does not position the wood against the side of the conveyor trough which is in line with the side anvil. Like most things, vibrating conveyors are not an undivided blessing. Their disadvantage is their low feed rate, which at best will not exceed 100 feet per minute. This low feed rate may require an increase in chipper size.

Undoubtedly, the single most difficult material to chip is kiln-dried trim ends. Because of their hardness, they have a tendency to explode as the knife hits the wood. Reduced disc speed and the use of vibrating conveyors with horizontal-feed chippers can nevertheless provide good results with minimum losses due to fines.

In British Columbia, there has been a trend to longer chips. I believe this is an attempt to reduce fines and therefore waste. Unfortunately, the cutting of longer chips requires an increase in the gap between the anvil and the disc surface (Figure 9.1). An increased gap will allow more pieces to be cut into "Texas chips" or fall through without being chipped at all. This is particularly true in gravity-feed chippers where the friction between the disc and the wood (shown by the arrow), helped by gravity, will increase this tendency. Perhaps a better way to reduce fines is to keep the chipper in top shape and reduce the cutting speed of the knives by slowing down the disc rpm. This will, however, reduce machine capacity as well.

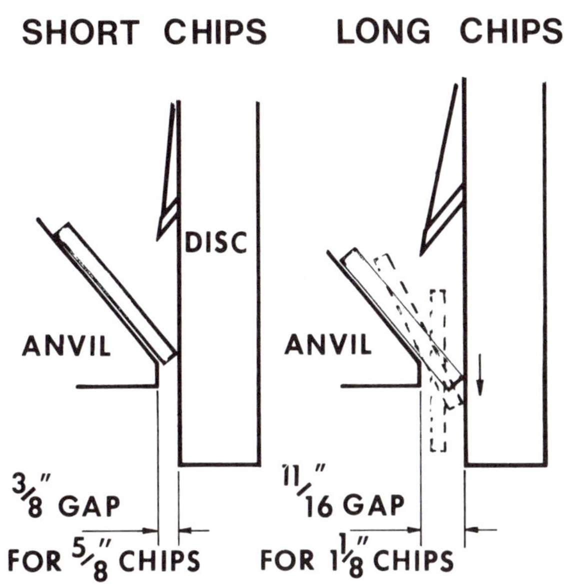

Figure 9.1. A longer chip length demands an increase in the gap between disc and anvil, allowing slivers and shims to fall through without being chipped. This is particularly true in gravity-feed chippers. "Overs" percentage will increase.

Probably the single biggest problem in chipping is the use of too small a machine. This causes more pieces to jam, thus requiring more men to keep it operating and resulting in headrig downtime.

Choose discharge type carefully

A final point is the selection of discharge. Should you use a top or a bottom discharge? A top-discharge chipper (Figure 9.2) has the initial advantage of easy installation. But look ahead before you decide. A chipper is designed to chip, not to blow. Blowing is at best a compromise proposition and an inefficient use of energy. Air movement is noisy. Unless the cyclone is big enough, you may get air pollution. Blowing distance is limited. It can be increased by relying more on "throwing." This, however, will not improve

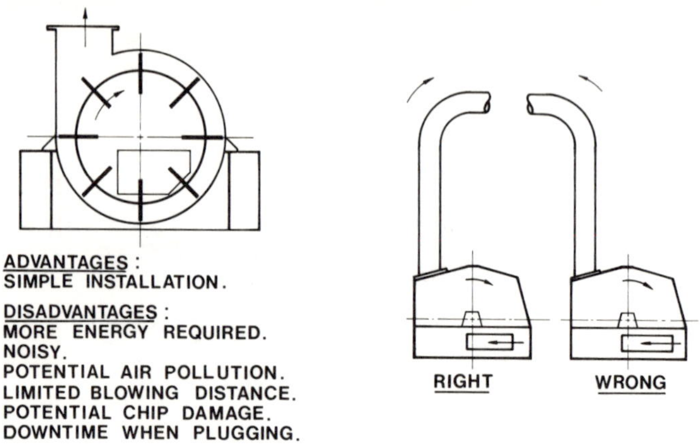

Figure 9.2. *Left:* Disadvantages of a top-discharge chipper may outweigh advantages. *Right:* To minimize chip damage in a top-discharge chipper, proper duct design is important.

chip quality! If you wish to reduce the chipper disc speed to improve chip quality, the speed may no longer be adequate for satisfactory blowing.

If you still decide to use a top-discharge chipper, be sure to design the ducting properly. Most of the chips travel on the outside of the duct. So avoid changes in direction which cause impact. Impact deteriorates chip quality and increases the tendency to plug.

Chip quality a product of careful design

So much for basic selection of size and type of chipper. Let us turn to some specific chipper design features which you should look for to produce the best chip quality. Generally, a machine producing good-quality chips will not suffer from frequent breakdowns. This is not to say that good preventive maintenance should not be carried out.

Although not fail-safe, it is a good bet that the heaviest chipper for a given spout size will provide the best chip quality, reduce maintenance, lessen downtime due to breakdowns, and have a longer life.

Chipper knives pass the anvils at a speed of about 90 miles per hour. As many as 5,000 knife cuts may be made per minute at this speed. Obviously this requires a robust, yet precision, machine.

Good chip quality can only be achieved with knife-to-anvil clearance at a minimum. Most of us have tried at one time or another to cut heavy paper or cardboard with poor scissors with a gap between the blades. The cardboard will bend or tear. A chipper has much in common with a pair of scissors.

Knife-to-anvil clearance is affected by several things. One is disc and arbor design (Figure 9.3). As the distance between the knife pockets on the disc decreases toward the center, the theoretical disc design would be tapered to make up in thickness what is lost in width. The arbor should be heavy and bearings should be mounted on both sides to improve the rigidity of both the arbor and the disc. This allows the knives to run closer to the anvils without ever hitting them. While it is not practical to produce a disc with a tapered back, rigidity of the disc is much improved by using a cast steel disc which may be cast thicker in the center than at the edges.

During chipping, several forces occur which will cause the disc to lift away from the base. Solid arbor bearings provide a more rigid disc assembly and reduce maintenance.

Of course, a rigid disc and arbor and a heavy base are of little value unless the knives are properly clamped. When thin, clamped

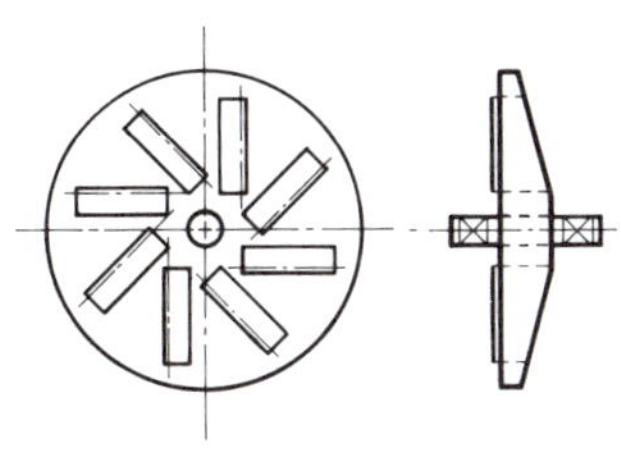

Figure 9.3. Ideal disc and arbor design includes tapering the back of the disc for constant strength between knife pockets and maximum rigidity. Bearings should be placed on both sides and close to disc.

Top: To improve chipper disc rigidity, use a cast steel disc which is thicker in the center than at the edges. *Bottom:* Solid arbor bearings improve disc assembly rigidity and reduce maintenance.

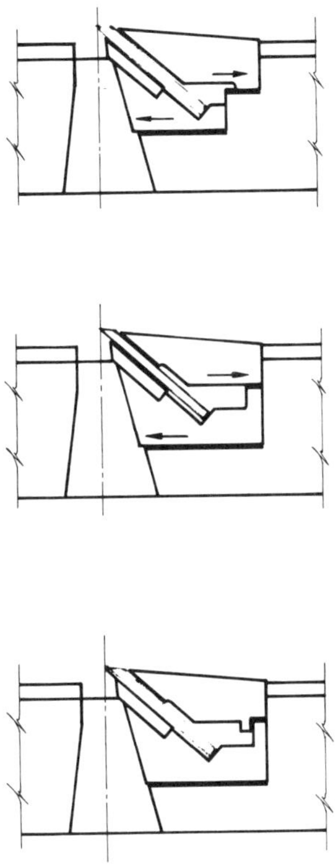

Figure 9.4. *Top:* In early chipper knife clamp design, clamp rested on ledge in disc. *Center:* Clamp rests on carrier in one improved design. *Bottom:* Projection under clamp eliminates horizontal sliding of knife and is the best design.

knives were first introduced, they looked like the top drawing in Figure 9.4. The clamp rested on a ledge in the disc. This meant the clamping action did not take place independent of the disc. Because of the angle in which the knife is seated, a horizontal force occurs, causing the clamp and knife carrier to open up so sawdust gets packed in and chip quality is reduced. At worst, the knife would hit the anvil and cause considerable damage.

The center picture shows an improved design where the clamp rests on the carrier; clamping action is thus independent of the disc. If wear takes place, the disc does not have to be machined or replaced, at least not in the short run.

The bottom drawing in Figure 9.4 shows an improved design whereby a projection under the clamp eliminates any horizontal sliding due to the knife angle. The cross-sections of the clamp and carrier should be sufficient to avoid undue deflection, which will cause the gap at the knife to open up. Clamping must be done with a preset torque wrench.

Trying to cut something with scissors with the blades wide open will cause whatever you are cutting to slide away, resulting in a poor cut. The same applies in a chipper spout. The wood should have stable support against the anvils, which form one blade of the "scissor." Most wastewood has sawn faces at right angles. For stable support, the anvils should therefore also be mounted at right angles to each other.

In horizontal-feed chippers, most chipping takes place against the side anvil (Figure 9.5). A wide bottom anvil might look good to you, but consider what happens when a board wider than the side anvil is fed. The board will be pushed by the knives toward

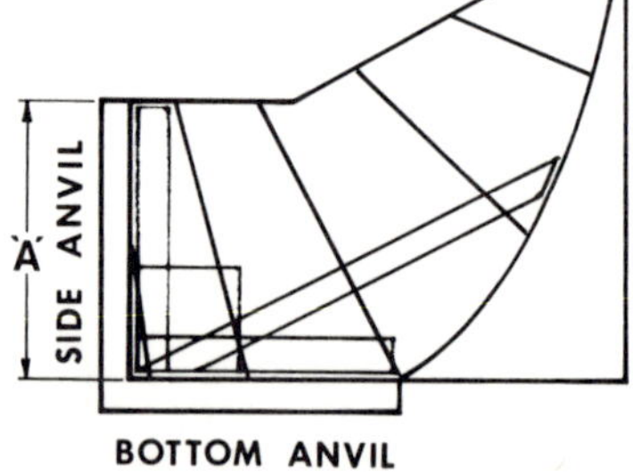

Figure 9.5. To provide stable support during chipping, and thus optimum chip quality, in all horizontal-feed chippers the wood should be narrower than the side anvil (A).

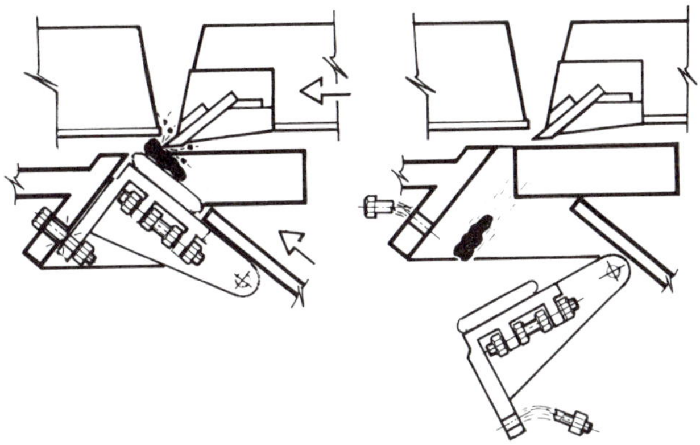

Figure 9.6. A pivoted side anvil protects against tramp metal.

the side anvil. But because it is wider than the anvil, it will get stuck and be torn by the knives, as it is not supported. If you want good chip quality, this is definitely undesirable.

The knives must "sweep" through the spout opening, forcing the wood into the corner but without uplifting the wood. Should the wood be lifted, chip length will be more variable and, at worst, the wood will be rejected and thrown back by the knives.

Chipper should have protection

A last point is protecting the chipper against damage from tramp metal. By far the best insurance is a reliable metal detector. Human nature being what it is, however, a detector may not be foolproof.

Consider the consequences of a bent arbor, a bent disc, or ruined knife clamps and carriers. It is worthwhile to invest in additional protection against tramp metal or rock damage. Considering 5,000 knife cuts per minute, or nearly 100 per second, preventive action has to take place quickly. Our method is to pivot the side anvil and hold it in place with shear bolts (Figure 9.6). As soon as the pressure on the anvil becomes excessive, the shear

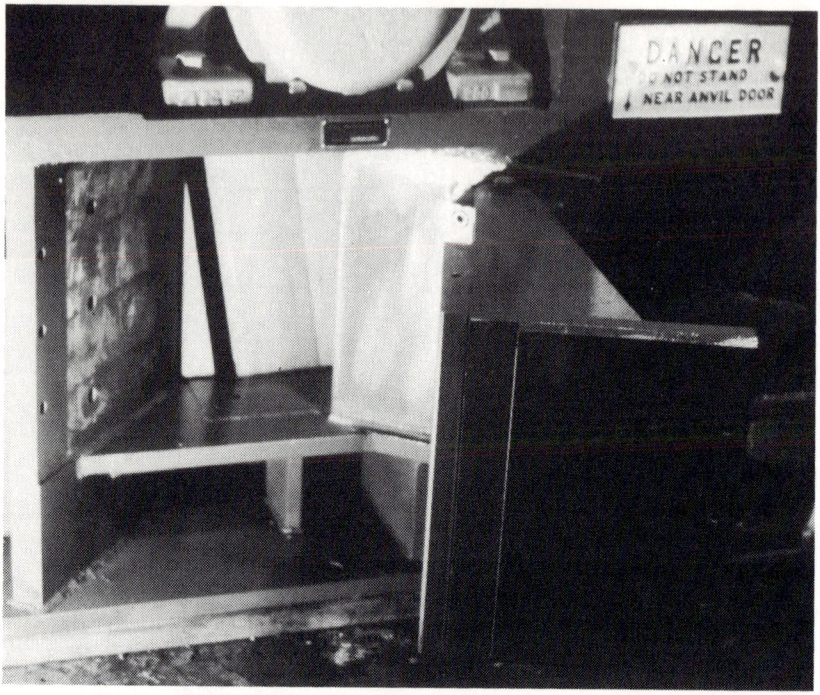

Shear bolts used to hold pivoted side anvil in place will let go when pressure from tramp metal or rocks becomes excessive. This protects chipper from damage and also offers easy access to both anvils.

bolts let go and tramp metal or rocks are discharged through an opening in the side of the spout. This method has proved very effective and has saved many a mill thousands of dollars in repair and downtime costs. This method also provides easy access to both side and bottom anvil, allowing easy inspection and adjustment.

I have tried to convince you that the chipper is an important machine in your mill operation. It can contribute considerably to mill revenue, providing the right machine is chosen for your particular operation. Unlike the well-known statement applied to our electronic aids, chippers, well-chosen and well-maintained, are definitely a case of GIGO: garbage in but *gold out.*

10

Improving drive efficiency with synthetic flat belts

Greg Baird, Northwest Operations Manager,
Habasit Belting Inc., Seattle, Washington

The subject is belting for power transmission. Let us review the principles which explain the superiority of some belts over others and why industry is beginning a return to the use of flat belting for all sizes and types of belt transmission drives. I will confine comment to power transmission, although the principles apply equally to a majority of plant lumber and plywood conveying problems.

Certain basic rules must be adhered to to get the most out of a belt drive. Knowing what to look for in a drive, using a set of known criteria, is a perplexing problem.

What are the basics to keep in mind about belting? Can we expect all belting suppliers to adhere to them? The answer to the latter question is both yes and no. Certain principles are compromised for various reasons. This is done with full knowledge. Some compromises are justified. But the principles that govern *all* belts are these:

1. High coefficient of friction.
2. Balanced elasticity.
3. Sufficient tensile strength to pull.
4. Low unit weight.

5. Low incidence of induced vibration.

All influenced by the principle of effective pull

$$\frac{T_1}{T_2} = e^{\mu a}$$

Whether the belt is trapped in a groove or running flat on a pulley, these characteristics must be present in the belt material in proper relationship to each other to deliver the most power for the least unit cost. The basic belt equation $\frac{T_1}{T_2} = e^{\mu a}$ has an overriding influence on them.

Constant coefficient of friction the goal

For example, you should expect the coefficient of friction to be constant throughout the speed range. It should be as high as possible without causing excessive belt wear and it should be predictable and made known. Not all belt materials meet this requirement. Few belting suppliers offer these data to you. The attitude seems to be, what difference does it make as long as the drive turns the wheels.

Very few V-belt supplies tell their customers what coefficient of friction they use. The industry standard is that if the belt is properly tensioned, the coefficient of friction will be 1.5, but that depends on tension and it depends on the length of time the V-belt drive has been in use because the constant rubbing in and out of the groove changes the coefficient of friction.

Figure 10.1 shows a constant coefficient of friction from 0 to maximum, which is what we want. With a leather belt (II), the coefficient of friction varies with belt speed, so it starts low (0.2) and builds up to 0.6 at 6,000 or 7,000 feet per minute, then begins to flatten out and drop off. Unless we want slip, a varying coefficient of friction is not a good feature.

An examination of belt drive principles will show that the inefficiency of some drives costs you money. Consider totally synthetic flat belting as one alternative. In the mid-1960s, when I was asked to come up with a drive, it was a simple matter of reaching for the Dodge catalog, opening it, reading the charts and graphs, and coming up with some pulleys and some belts. That was the

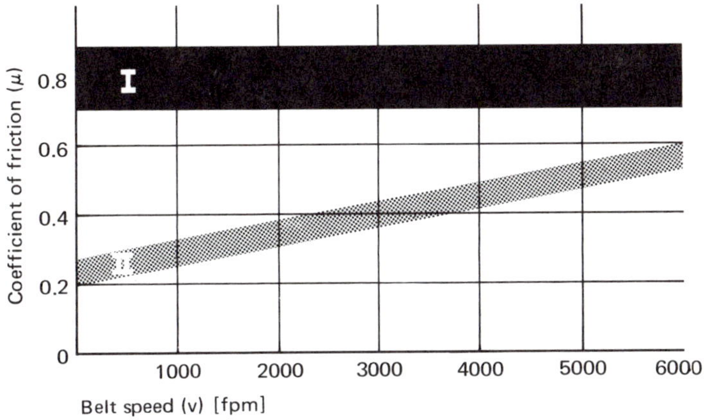

Figure 10.1. Coefficient of friction should remain constant with increases in belt speed. All-synthetic flat belt (I) meets this criterion. Leather belt (II) does not.

last I saw or heard of the drive. Since then, however, I have seen a tremendous number of mill problems all related to belt drives.

Synthetics introduced to belting in 1920s

To digress briefly, history records that belting has existed for 500 years. Until about 50 years ago, however, all belts were made from natural materials. Leather, cotton, cotton with balata gum, and hemp are a few examples. To attempt to meet the needs of modern machinery, these natural materials were combined with man-made materials 50 years ago, allowing belt manufacturers to predict with greater certainty the performance a belt would offer. At the same time, cotton-cord-with-rubber V-belts were introduced in the automotive industry. This combination was produced in an attempt to improve the properties of the natural materials then known to everyone. In the beginning V-belts were intended for use on small horsepower drives like fans and automotive engines—not as replacements for flat belting. But because flat belts remained inadequate for so many years, V-belts gained greater prominence.

Natural materials never quite measured up in many ways. It was thought that by adding predictable, man-made elements to these marginally adequate materials, a superior belt would result. Synthetics offered a crutch to the natural materials, but the combinations are only marginally successful even today. Yet combinations such as leather with nylon or polyester, however inefficient, are still in use.

It was not until totally synthetic, predictable materials came on the scene in the post-World War II years that any significant changes took place in belting. We began to see how we could balance the properties of a belt to meet the demands imposed. But progress continues, and we see a trend for the future: that trend is a return to flat belt drives, because the all-synthetic flat belts available today meet the criteria.

Elasticity and tensile strength must be balanced

A belt must be "elastic" but not "plastic." It must absorb shock without permanent stretch, which merely means elasticity and tensile strength are balanced. We are looking for belting material which has a predictability property that relates tension to pulling power. If you can accurately determine the additional horsepower a belt will deliver with each percentage of tension added, you have the properties for a versatile, efficient drive.

If you cannot predict this—and you cannot with V-belts—the belting does not meet the requirement. The equation for effective pull $(\frac{T_1}{T_2})$ influences the properties which give us this balanced material.

Figure 10.2 illustrates shock absorption. A peak impulse occurs when a shock is applied to a running belt. If the belt is not elastic but a fairly rigid material (a steel band, for example), the response will be as shown by F in the illustration. If the belt is properly balanced elasticity-wise, it will spread out the shock over time, as shown by F_R. It will flatten it out, lowering the peak. The energy—that is, the area under each peak—is the same in both cases. It is simply spread out a bit more in F_R. We are looking for belts that will spread out shock and that will transmit very little

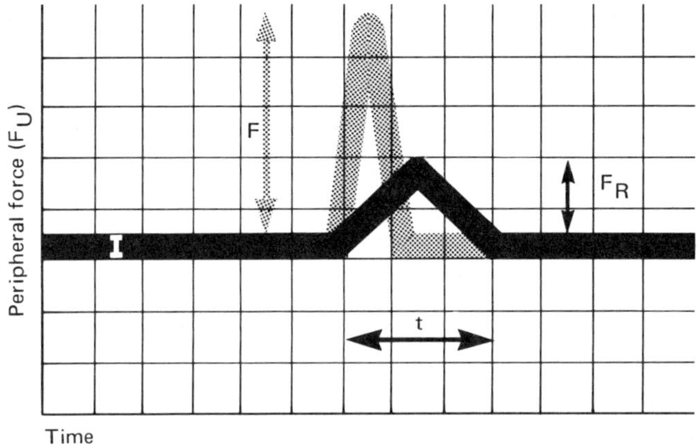

Figure 10.2. Belt elasticity is important for shock absorption. With a rigid belt material (*F*), impact creates a peak shock. In a belt with balanced elasticity (F_R), shock is spread out over time, protecting the motor and extending belt life.

back into the motor. V-belts are pretty good in this respect. In fact, V-belt manufacturers advertise that their belts absorb shock. They do absorb quite a bit of shock, but not enough to preserve the life of the belt and to prevent belt damage.

The property of high tensile strength tells us very little by itself. A steel belt has great tensile strength, but depending on what is used to create the necessary grip, it is a rather unpredictable drive. By the same measure, tensile strength in pounds-per-inch working load tells us nothing about the other characteristics.

Examining the bearing-load question

Bearing load is of great concern when flat belting is discussed. The extent of the bearing load is governed by a proper balance of coefficient of friction, elasticity, tensile strength, and unit weight of the material. These are controlled by the ratio of T_1 to T_2. Comparing a leather-faced nylon belt, the direct equivalent in an all-synthetic belt, and a V-belt (Figure 10.3), we see the effect on bearing load caused by lower elasticity and higher unit weight. The centrifugal force must be compensated by higher dynamic tension. Here *balanced* tensile strength and proper elasticity come into play.

Curve No. 1 in Figure 10.3 represents an all-synthetic belt with fairly low unit weight. The effect of centrifugal force (the force that causes the belt to leave the pulley and that is dependent on weight, which is WR^2 forces) is the least, of the three belts compared. Since leather and nylon (curve No. 2) weigh about the same per cubic inch (0.035 to 0.040 pound per cubic inch), a nylon belt with a layer of leather will give you a belt about double the weight of a nylon-only belt. V-belts are shown in curve No. 3. The fact than an 8V V-belt is such a large mass of rubber is reflected in the effect of speed. It is very pronounced when compared with a very thin, high-strength synthetic belt.

Figure 10.4 shows a comparison of peripheral force with an all-synthetic nylon belt (A-3) and two V-belts. The designations are for European belts. The SPA is a special A-belt, which would

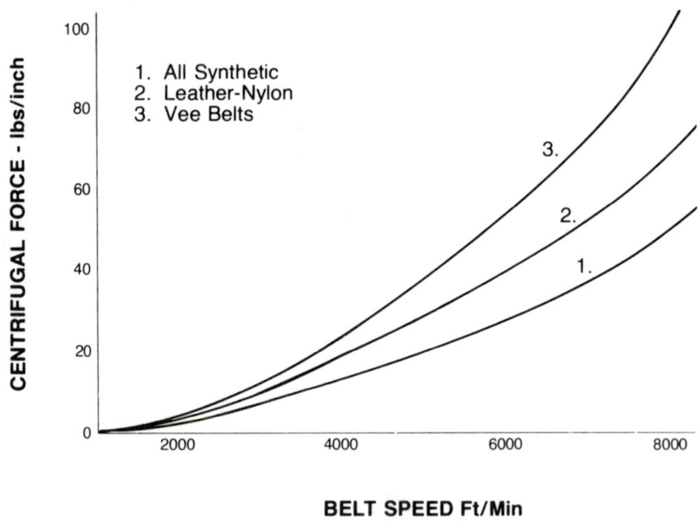

Figure 10.3. Effect of centrifugal force and its relation to belt speed for three types of belts. Effect of centrifugal force is least with light-weight all-synthetic flat belt.

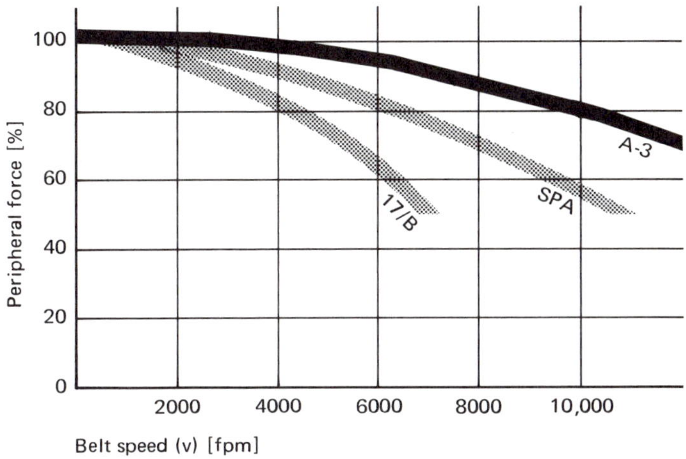

Figure 10.4. Horsepower delivered drops significantly above 5,500 feet per minute with V-belts (SPA and 17/B). All-synthetic flat belts (A-3) allow belt speeds to be raised for greater efficiency.

be similar to a 5V; the 17/B is a B-section V-belt. The horsepower delivered by the V-belts drops off very quickly above 6,000 feet per minute. Of course, V-belts are not designed to be used above 5,500 or 6,000 feet per minute. Thus, they are very restrictive because belt speeds cannot be increased to the point at which they become really efficient. A flat belt, however, becomes very efficient above 6,000 feet per minute. It can pull as much as 120 horsepower per inch of width at about 10,000 feet per minute. So for edger drives and other drives where pulley diameters and belt speed can be increased, a 9-inch flat belt can pull 1,000 horsepower. (And the tension is not much greater.)

Figure 10.5 compares bearing loads for Optibelts, a European brand of V-belt, and all-synthetic flat belts. Bearing load is about the same throughout the chart. V-belts are a little higher in the 60 to 75 horsepower range. Flat belts are a little higher—2,500 pounds for flat belts as compared with 2,200 pounds for V-belts at 175 horsepower—in the 125 to 175 horsepower range. But if bearing load is critical—plus or minus a few hundred pounds—then there is a problem regardless of belt drive selected.

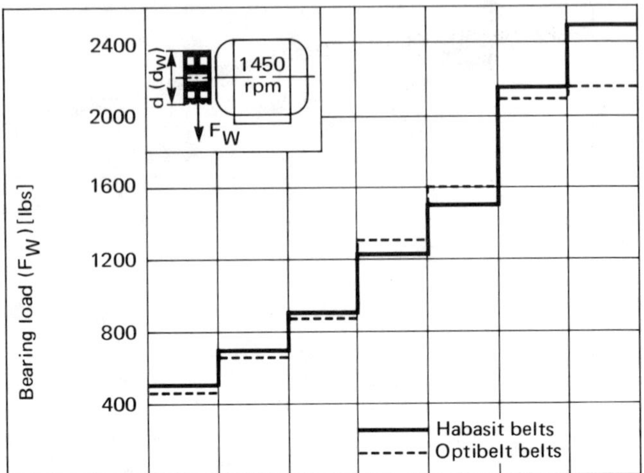

Motor type	160 M	180 M	200 L	225 M	250 M	280 M	315 M
Motor output	15 HP	25 HP	40 HP	60 HP	75 HP	125 HP	175 HP
Habasit belts	A-3			A-4			A-5
Pulley dia. (d)	6.3 in.	7.1 in.	9 in.	10 in.		11.2 in.	16 in.
Belt width (b_o)	2.16 in.	3.0 in.	3.5 in.		4.3 in.	6 in.	4.13 in.
Optibelt belts	SPZ	SPA		SPB			
Pulley dia. (d_W)	6.3 in.	7.1 in.	9 in.			11.2 in.	16 in.
Number of belts	3		4		5	6	

Figure 10.5. Comparison of bearing loads for V-belts (Optibelt) and for all-synthetic flat belts (Habasit).

Effects of vibration examined

Induced vibration has always been a consideration in belting. It has never been desirable. Years ago, vibration was not a serious problem. Speeds were much slower, and a well-spliced flat belt did not induce appreciable vibration. By the same token, V-belting creates considerable vibration. And, we have not eliminated the problem in recent years; we have merely beefed up the motor bases and the equipment to dampen it. In the past five years, reduced elasticity in V-belts, in conjunction with lighter frame electric motors, has compounded the problem.

Recent data indicate the tremendous effect of vibration on bearings. I have seen bearings destroyed in six months from 1 mil of axial vibration. I have heard it said that V-belt pulleys do not need to be lined up as well as flat-belt pulleys. I can tell you it is more critical that V-belts be lined up properly. First, if a flat belt is not lined up, it will not run at all. A V-belt *will* run. Correction of misaligned drives would eliminate many machinery problems.

There is a distinct relationship between misalignment and the center distance of the drive. Buy a different length V-belt—go a little longer and move the motor back a little. That usually solves the problem. We have also found that 5V V-belts seem to be the worst culprits in terms of vibration.

Most vibration problems are caused by the splice on the V-belt. On a typical V-belt there is a 0.020- to 0.030-inch difference in thickness in the splice. The belt is trapped in a groove. When the splice comes around, it must cause something to happen, because all the power is taken off the edges of the groove. What happens? It vibrates axially, because that is where the forces are.

To give you a graphic example: If you manufacture 500,000 board feet of lumber per year and you have 10 ripsaws in the mill, it probably costs you about $600 per year for every 0.005 inch those saws vibrate while in the cut. Our data show that saws vibrate more than 0.005 inch and that axial vibration is the culprit. The drive may be only part of the problem, but evidence leads us to believe it is a considerable part.

The predominant cause of axial vibration in bandmills, edgers, fans, and chippers is the V-belt drive. We have tested old bandmills with flat bearings and out-of-balance wheels and have isolated the problem to the drive.

That covers the basics. Other secondary principles apply in this discussion. Some are important in sawmills. Antistatic belting is a good example. The ability to make good splices in the field is another. All-synthetic flat belting can be spliced in the field. It is very versatile. You can put the motor where you want it; you are not tied to a particular length. We have been talking recently about flat-belt drives for merchandisers. Saw sticking is a problem—with a line shaft drive, all the power in the motor goes into the one belt. V-belts slip and burn up. There are cases where flat belts are not the answer, or where other solutions, such as overriding clutches, can compensate for saw binding. Yet efficiency- and vibration-wise, all-synthetic flat belts are superior to V-belts.

Table 10.1. Comparison of All-Synthetic Flat Belt and Leather-Nylon Belt for Edger Drive

DATA
MACHINE: CIRCULAR GANG (Edger)
MOTOR: 500 HP @ 1800 RPM
DRIVE RATIO: 1.1 : 1 (Speed Up)

	DIRECT COMPARISON		RE-DESIGN
BELT TYPE	LT-40	A-4	A-5
CENTER DISTANCE (Inches)	78	78	78
BELT SPEED (FPM)	10,700	10,700	10,700
DRIVER PULLEY DIAMETER (Inches)	22¾	22¾	22¾
DRIVEN PULLEY DIAMETER (Inches)	20	20	20
DESIGN HORSEPOWER	800	800	800
BELT WIDTH (Inches)	11	10	7
DYNAMIC SHAFT LOAD (Pounds)	6,200	5,000	5,200
DRIVE DESIGN S.F.	1.6	1.6	1.6

Belt drive types compared

Now, I would like to compare all-synthetic flat belts, combination belts, and V-belts in three typical drive situations.

Table 10.1 shows a circular gang edger with an all-synthetic flat belt drive (A-4); I have compared a leather-nylon material (LT-40) with the all-synthetic belt and then I have shown what I did to improve the drive in the "redesign" column. This edger has 500 horsepower on each arbor; we have designed the arbor horsepower for 800. The center distance is 78 inches. We are running the belt at 10,700 feet per minute, which is quite a bit higher than a V-belt drive. In fact, the mill had been driving the edger with an 8V at about 5,500 or 6,000 feet per minute. Notice the dynamic shaft load: the leather-nylon belt, with a lower coefficient of friction, creates a higher bearing load–6,200 pounds. The all-synthetic belt creates a bearing load of 5,000 pounds. In the redesign, we used a 7-inch-wide belt running on the same pulley diameters–22.75 inches and 20 inches–to create a 5,200-pound bearing load. Of course, bearing load was not a problem here at all.

Table 10.2 shows a chipper, which is still driven by an 8V V-belt. They have some problems. In fact, they have taken off half the V-belts and it still runs perfectly well but vibration is less so problems have been reduced greatly. A direct comparison between the 8V now in use and the all-synthetic belt (A-5) shows these differences: They have designed it for 1,500 horsepower. We do not need that. We can give them 1,200 (which is 1.5 over 8) and we can guarantee 1,200 horsepower. Notice the bearing loads. Static shaft load is 14,000 pounds with an 8V drive and 14,800 pounds with an all-synthetic flat-belt drive, so the difference is only 800 pounds. What we would propose, however, is to change the pulley diameters a bit, raise the belt speed for greater efficiency, and narrow the belt to a 14-inch width. That would give a full 1,200 horsepower with a medium bearing load for that size drive—it is a big chipper.

Table 10.3 is a typical drive for a bandmill. The A-5 all-synthetic belt has been in use on the 5-foot Letson & Burpee high-strain mill (100-horsepower motors) for two years. Originally Letson & Burpee proposed a 5V drive, but the owner bought it with a flat-belt drive instead. Note the static shaft load. Had it

Table 10.2. Comparison of All-Synthetic Flat Belt and High-Duty V-Belt for Chipper Drive

DATA
MACHINE: CHIPPER
MOTOR: 800 HP @ 1200 RPM
DRIVE RATIO: 3.77 : 1

	DIRECT COMPARISON		RE-DESIGN
BELT TYPE	8V	A-5	A-5
CENTER DISTANCE (Inches)	104	104	104
BELT SPEED (FPM)	5975	5975	6290
DRIVER PULLEY DIAMETER (Inches)	19	19	20
DRIVEN PULLEY DIAMETER (Inches)	71	71	75
DESIGN HORSEPOWER	1500	1200	1200
NUMBER OF BELTS OR BELT WIDTH	16	15	14
STATIC SHAFT LOAD (Lbs.)	14,000	14,800	14,800
DRIVE DESIGN S.F. (From Catalogs)	2	1.5	1.5

REF. GATES DRIVE DESIGN MANUAL

Table 10.3. Comparison of All-Synthetic Flat Belt and High-Duty V-Belt for High-Strain Bandmill Drive

DATA
MACHINE: HIGH STRAIN BAND MILL
MOTOR: 100 HP @ 1200 RPM
RATIO OF DRIVE: 2.13 : 1

	DIRECT COMPARISON		RE-DESIGN
BELT TYPE	5V	A-5	A-5
CENTER DISTANCE (Inches)	50	50	50
BELT SPEED (FPM)	4875	4875	5660
DRIVER PULLEY DIAMETER (Inches)	15.5	15.5	18
DRIVEN PULLEY DIAMETER (Inches)	33	33	38
DESIGN HORSEPOWER	175	175	175
NUMBER OF BELTS OR WIDTH	5	3	2.5
STATIC SHAFT LOAD (Lbs.)	1900	2200	1900
DRIVE DESIGN S.F. (From Catalogs)	1.5	1.5	1.5

REF. GATES DRIVE DESIGN CATALOG

come in with a V-belt drive, static shaft load would have been 1,900 pounds. Our direct comparison check is 2,200. What I would rather do, however, is raise the belt speed a little, increase the diameter of the pulleys (keeping the ratio the same), and bring the static shaft load to 1,900 pounds, which is exactly the same as the V-belt loading. I hope these tables help dispel the tale that flat belts create huge loads. These examples show it is not the case.

The primary purpose of information I have given is this: to emphasize the potential of flat-belt drives in the sawmill industry. Increased production, power savings, longer belt life, and reduction of vibration are the economic consideratons which must be clarified no matter what belt drive you select. Habasit is the all-synthetic belt I have been discussing. We make the only totally synthetic, complete line of transmission and conveyor belting. We are a worldwide company, based in Basel, Switzerland.

Author's note: I want to acknowledge technical information supplied by Walter C. Bass, president, Habasit Belting Inc., Habasit AG, Basel, Switzerland, and my friends within the industry who collaborated with me in gathering data.

DISCUSSION PERIOD

QUESTION: How efficient or inefficient are V-belts with a power band back?

BAIRD: That is a good question. The backing compounds the problem of mass or weight. The belt is heavier by virtue of the heavier band of rubber. You do not get the typical problem of rollover, of course. But vibration problems are worse with bands, elasticity is sometimes reduced, and they are really not nearly as efficient in terms of horsepower delivered. Efficiency of banded V-belts is 88 percent, as compared with 98 or 99 percent for a good, high-friction flat belt.

11

Building a practical quality control program

Dean Huber, State & Private Forestry, Cooperative Forestry, U.S. Forest Service-Region V, San Francisco, California

George H. Sisterhenm, Project Manager, Forest Products Group, Fibreboard Corp., San Francisco, California

Robert L. Mardock, Operations Manager, Harwood Products, Branscomb, California

James L. Russell, Quality Control Manager, Idaho Forest Industries, Coeur d'Alene, Idaho

Robert W. Nix, Utilization Supervisor, Diamond International Corp., Coeur d'Alene, Idaho

Frank John Banducci, Jr., Quality Control Supervisor, Flakeboard Operation, Roseburg Lumber Co., Roseburg, Oregon

Introduction by Dean Huber

Mr. Manager, I have a question for you: "Are you a *thermometer* or are you a *thermostat?*"

If you are a thermometer, you will simply respond to the conditions around you. You will fluctuate from hot to cold to hot. Your actions will be erratic, with no means of control.

If you are a thermostat, you will have a means for control. You will be working toward a defined target and responding within established limits. You will be striving to maintain a stabilized system of production, throughput, product quality, and a sensible system for profit and cash flow.

What, then, provides the method for the mill manager with control? My suggestion—and that of the others represented here—is to establish a practical quality control program.

Mr. Mill Owner, Mill Manager, Mill Operator—you are the one who establishes policy and makes the decisions. Those decisions are passed down the line through the production system and eventually evolve as a product. But just how effective is your system? How do you know what is really happening in your operation? What do you have for feedback from the production line and the customer? Is your operation set up to function with an action-and-response system?

I have a suggestion. *Establish a quality control program* and put the responsibility for that program into the hands of a capable person. Five individuals will discuss this person and his program. Each will provide a building block in the formulation of a quality control program. These men have an important message to convey. I guarantee it is not a discussion of motherhood and apple pie. It stresses the practical side of quality control:

Where does it start?
What does it involve?
Who does it involve?
How broad is its application?
What can it do for you?
Should *you* have a quality control program?

How management sets the framework for quality control
by George H. Sisterhenm

In recent years, the forest products industry has experienced a rapid and staggering increase in labor and log costs. These increases have been offset in part by large capital expenditures for labor-saving devices and by process changes that have increased fiber yield. Process changes must achieve more complete utilization of the basic raw material, the log.

While some managers have taken this approach, others have gone back to the basics of mill operation—that is, stressing fundamental management procedures. They have also taken a new look at their human resource as their prime asset. This approach often involves low-cost capital programs, one of which we consider quality control.

Most mills have some form of quality control. You may not identify with the programs we will discuss, as your program may not be quite so sophisticated. In the past, quality control programs have demanded a great deal of accountability. In so doing, the mention of quality control has generated a considerable amount of friction and animosity at all levels of management. In fact, some of you may have rather bitter memories of your problems with quality control men.

Our objective is to show you the favorable aspects of quality control and to point out the cautions involved in developing or expanding a program. The information I will offer is based on an industry survey consisting of my observations and those of others as to the state of the art in quality control today. It is not theoretical, but reflects existing practices.

Buyer demands quality

Where are we today and what are we facing? Today is the reign of the consumer advocate. Increasing consumer activity stresses the need for quality in all products. This has created pressure on contractors, local building code officials, trade associations, and finally the manufacturer. Consumer complaints are on the increase and will continue to do so even after business improves. The buyer wants the performance he has paid for and we must be sure he gets it. Emphasis must be placed on quality control.

You can call this program a "quality control program," "quality assurance," or "process control." I prefer the designation "process control" or "production control." These terms denote the full range and responsibility of the program. Remember, as managers we are not thinking only about quality. We may have to alter product mix, log allocation, and manpower to achieve the ultimate goal—maximizing profits through the best fit of man, machine, and raw material. Our quality control program goes from the stump to the deck to the mill infeed to the boxcar. It is designed as a complete quality control program and involves all phases in the conversion from logs to lumber. The program should target not only one or two areas, but rather should use the shotgun approach for broader coverage of the entire operation.

I purposely have omitted a discussion of the specific costs of a program and the anticipated dollar gains. Why? Because each mill is different. Mills vary in size and production capacity. The scope of application of the program will vary and costs will vary. The basic procedures of statistical quality control have existed in other industries for years. There is nothing secret about them. We have progressed in the lumber business from lumber calipers in 1/32-inch increments to dial calipers in 0.001-inch increments. We have stopped talking about sawing variations of plus or minus 1/4 inch and are addressing ourselves to a standard deviation of 0.030 inch.

In-plant programs can range in size from one man with a caliper to a sophisticated statistical staff with computerized assistance. In some cases, certified testing agencies may be the entire quality control program in a plant. Ideally they should only complement an existing in-plant program. You may exercise more control over your plant quality control personnel, but you do need an outside source for purposes of comparison. You may not be able to afford the luxury of being the best, but with process and quality control, you may be one of the more profitable.

Elements of quality control program

What makes up a basic quality control program? Essentially, it is a data collection system with calculations, report processing, and distribution of the results. Programs contain all these things to a greater or lesser degree. But action and follow-up are the all-important parts of the program.

That brings in the area of communication. Without management support and the requirement for communication, a quality control program becomes an exercise in futility. All data must be timely. You as managers want to know what happened today—and you want to know it exactly. In today's world, a one-man show cannot realistically cover all the areas he must. The manager needs help.

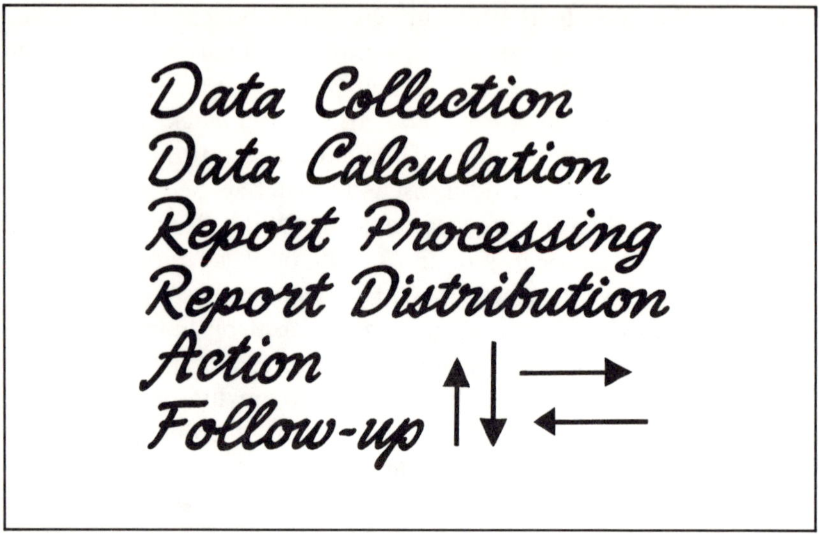

The basic elements of a quality control program.

The intent of an organized process or quality control program is to give him that help.

Where do we apply a quality control program? In the past, it has been applied primarily to headrigs and at the grading station, but today we are looking at all machine centers—resaws, sash gangs, edgers, the saw filing room. For instance, just knowing we have an excessive sawing variation as the board goes into the boxcar is not enough. We must know where it is occurring and correct it immediately.

Quality control data improve cost analysis

The data base that is developed in the quality control program—the numbers generated daily—fits into financial analysis by enabling a more accurately costing by machine center. We can then tie variations in green target thicknesses, grading efficiency, yield, or utilization to the operating statement. The effect of equipment and saw maintenance on the quality of the product produced is integral with production. The records we keep tell us not only what kind of a job we are doing and with what degree of control we are doing it, but they will accurately reflect the true cost. We can use this information, for example, to schedule saw changes or preventive maintenance as machine wear becomes noticeable—not necessarily in the machine, but in the product we produce. In other words, we will see the effect of machine or saw wear on paper before we actually see it or feel it in the boards.

Depending on the age of the mill and the quality of its maintenance program, these quality records may even indicate, with dollar justification, that *major* overhauls should be scheduled more than once a year. This can be a lifesaver to the maintenance supervisor, who never seems to get a machine down long enough to work on it—that is, unless it breaks down. Then it is generally patched, rather than fixed properly, before it is put back on line. This patching may help achieve a production goal, but the quality loss may do little for the profit goal.

Quality control reduces the frequency of judgment decisions. We would like to think it takes some of the guesswork out of the business. Quality control monitors control points when changes are made. When we alter sawing patterns, shift to thin saws, or regulate saw tensioning, we can measure board thicknesses, widths, grade, and check our numbers to see whether we have hurt or helped ourselves. The data collected should be sensitive enough to reflect these changes and then should be evaluated in terms of dollars and cents.

1- Pick the Right People
Pick the Right Program
2- Back it up
3- Use it
4- Up-date it

Keys to success with a quality control program.

You will find too that the quality control information collected will develop a data base for justifying plant expansion and/or process modification. This information will offer hard numbers for justification of larger investments and will assist in the post-audit evaluation after those investments have been made and the equipment is in and running. The more credibility we as operating managers develop in our handling of capital projects, the more money we will have to work with and the easier it will be to get.

Use top people in quality control program

In terms of people, select the best. You don't want a cull no one else wants. If there is someone in your mill who cannot do anything else, don't make him a quality control man. Use your best people. If you have a certified master grader, use him. If you have the best sawyer in the world, involve him. If as a manager you are not sure what kind of a quality control program you want or if you are trying to evaluate the program you have, professional help is available if you need it.

The manager must back up the program and the people involved with positive communication. Show that *you* use the program. Stress use of the quality control report along with the daily production report. They should go hand in hand. Your interest, whether as a manager or as a floor supervisor, will be infectious and will be reflected in the success of the program.

Update your program often. Emphasis can be selective as need dictates. For example, if you develop or learn of new testing techniques or of a better reporting system, keep your program current and updated. Change standards. Keep the program moving. It will take your support, as the manager, to make it work.

Believe me, quality control does increase profits. You will hear of specific examples later in this discussion. A quality control program will only work if you, the manager, want it to and if you put your full support behind it.

Now we know where to start—right at the top with management. Let us look now at where to apply a quality control program. This involves the machinery side of quality control and takes us into the grease pit, to the saw guides, and to the edger table. This is the grease-and-guts part of the program.

Applying quality control to the machinery

by Robert L. Mardock

I cannot stress enough how great an effect the monitoring of machinery has on quality control. This is where quality comes from. This is also where recovery comes from. You cannot increase recovery without quality control.

We feel quality control begins in the log yard. A log must be bucked properly for best recovery.

Quality control is needed at the headrig to get maximum recovery out of the log. You must have precision equipment to maintain good quality control. We now use an encoder box drive on our carriage knees. Originally the encoder was driven off the set shaft, but we could not maintain the tolerances set up by quality control with that drive. So we were ready to make the changes needed to maintain quality control at the headrig.

Left: Encoder box on headrig carriage is now driven by rack and pinion. Tolerances could not be maintained with original set shaft drive. *Right:* End view of carriage shows one attempt to eliminate set shaft torque. A second, larger set shaft was added below original one and direct-coupled to it with sprockets and chain. This was not a complete success.

Main control panel readouts can be monitored from both sides. At upper left is clock (1) to record carriage downtime. Below it is line counter (2) showing number of lines sawn per shift. Preset readout (3) and backstand readout (4) are at right in the photograph.

We now drive the encoder box with a rack and pinion. The original set shaft drive had too much set shaft torque. We did quite a bit of experimenting with different drives for the encoder to get better accuracy. Before we moved to the rack and pinion off the middle knee, we did try to eliminate the set shaft torque by adding a second set shaft. The original set shaft was a 2-15/16 inch. We added a 3-15/16 inch below it and direct-coupled it with sprockets and chain. This eliminated most, but not all, of the set shaft torque.

We have a Lloyd Controls setworks on our carriage. We did away with the inch wheel on the carriage. The sawyer does all his sawing directly from digital meter readouts. We have a preset, a backstand, and a backstand accumulation readout.

Our main control panel includes a number of data collectors which we feel are essential to quality control. The first is a clock to record carriage downtime. We also have a line counter to record the number of lines sawn per shift. All readouts can be monitored from the back side of the track by quality control and management. We use a proximity meter to monitor our headrig bandmill and measure saw deflection. This is very useful in maintaining quality control on the bandsaw.

Accuracy improved at edger

We have come up with several useful items for the edgerman through quality control. The use of laser lights is one of the biggest improvements we have made. We also have given the

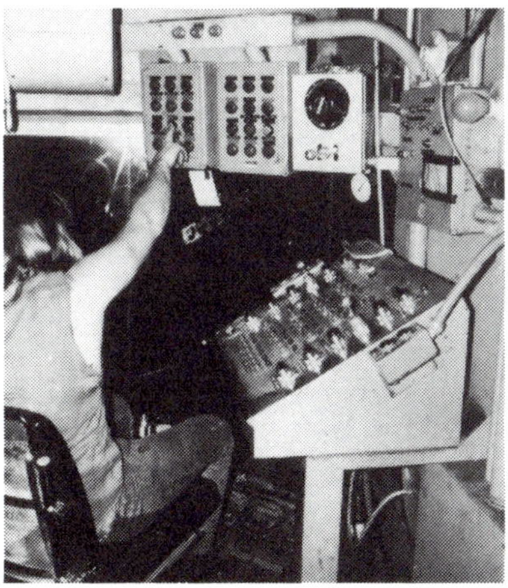

Left: Edgerman's control panel allows him to shift saw bank for better lineup of No. 1 saw. *Right:* Laser lights help edgerman improve recovery. This board, with crook, is lined up for best recovery.

edgerman tools to work with so he can improve recovery and gain better quality control. By monitoring the quality control charts, we know where to move on our edgers, our headrig, and in other parts of the mill so we can improve recovery.

We have devised a way to shift the bank of saws on our four-saw edger. A simple rack and pinion, driven by a gear reducer, moves the bank of saws to and away from the linebar for better lineup of the No. 1 saw. We found—again, through quality control—that by being able to crowd the linebar, by putting the saws in the wane edge where we need to have them, we could maintain better quality control and achieve better recovery.

We had the normal, run-of-the-mill cant flippers in our edger table. They were fine for Douglas fir, but with redwoods or other soft woods, we found we were losing quality. The cant flippers were gouging the cants or breaking them in half simply by turning them at the edger. We devised a scissors turner, which supports the cant through the entire turning process. This, again, was a product of our quality control program.

Our gang edger saws have an 0.085-inch plate with 0.125-inch kerf. We use an 8-inch splined arbor on the gang edger. We have done away with the collars on our circular resaw. I might add that on the gang edger we do not shift the bank of saws; we shift the linebar. We have a nominal preset and we also have a wild set on the linebar.

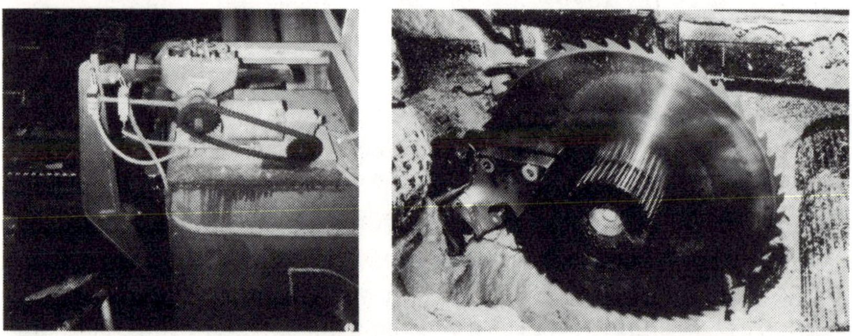

Left: Rack and pinion driven by a gear reducer moves bank of edger saws toward or away from the linebar for best No. 1 saw lineup. Being able to crowd the linebar, a modification suggested by quality control, has improved recovery significantly. *Right:* Gang edger saws have an 8-inch splined arbor. Collars are not used.

Scissors-type cant turners on edger table support redwood and other soft wood cants. When turned with conventional cant flippers, recovery was lost because cants broke in half or were gouged.

Our resaw is where we do most of our experimental work. We have a front guide, which is an identical copy of our gang edger guide, and we also have top guides. When the lid comes down, it holds one side of the guides. We are currently feeding the resaw at 500-plus feet per minute with a lumber thickness tolerance of 0.030 inch or less. We use different feed speeds for redwood than for fir. The 500-feet-per-minute speed is on redwood—and we have gone as high as 600 feet per minute with the same accuracy.

Graphs used to maintain control

All our lumber is checked by quality control using a dial caliper, in 0.001-inch increments. Our quality control man takes his samples, records them, and then puts them into graph form. At the end of the month, he compiles all his samples, puts them onto a graph, and then onto a data readout sheet, which shows at a glance the on-size, undersize, and oversize. Each machine has its own graph. At the end of the month, each floor foreman gets a copy of the graphs for all the machines, so he can visually see what they

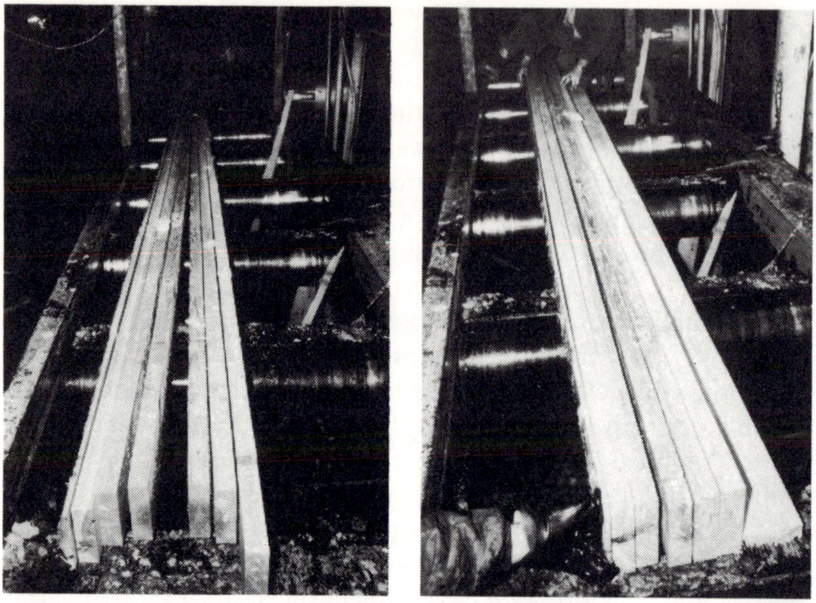

Lumber leaving gang edger, cutting 2x6s, shows importance of being able to put first saw in wane edge and crowd linebar. *Left:* No. 1 saw was not in best position. Board on right would normally have made a 2x6 but because of wane edge on lower part, it will only resaw to a 1x6. Had wood in slab at left been applied to board at right, recovery would have been 9 percent better. *Right:* Here No. 1 saw was placed and held where it belongs. Board at right will not make a 2x6, but it will make a 1x6. Had this cant been edged as in left-hand photograph, however, the 1x6 would have been lost—for a 10 percent loss in recovery.

have been doing. Also, when the quality control man comes into our mill, the standard procedure is for either the floor foreman or the millwright foreman to accompany him to the lumber pile or to the chain while he measures the lumber. The foreman knows immediately what quality control has seen. The floor foreman also has his own dial calipers which he packs with him. He monitors his own equipment daily, hourly, or by the minute.

The graphs make very versatile quality control charts because we can change our sizes at random. It is no problem to narrow them 0.020 or 0.030 inch, or do whatever we wish with them. We are now in the process of redefining our tolerances, because we have achieved the goals we set out to reach. Now we are going to tighten them up and reduce the allowable tolerances.

In summation, in a good quality control program, monitoring the machinery is the key to greater recovery. And for quality control to work, it must be backed by management.

Let us look at another building block—people. Both George Sisterhenm and Bob Mardock indicated that people are a vital part of a quality control program. That is exactly what Idaho Forest Inddustries found when it began a quality control program about four years ago. Theirs was a typical old-fashioned mill with old-fashioned people who had old-fashioned ideas. How do you implement a quality control program when you are starting with everything old? The key to their success: They did not give up. They kept trying. When it didn't work, they tried again. And very honestly, they are still trying. But they have come a long way with the people side of quality control.

A quality control man checks lumber with dial caliper. All lumber is checked by quality control and results are put into graph form monthly.

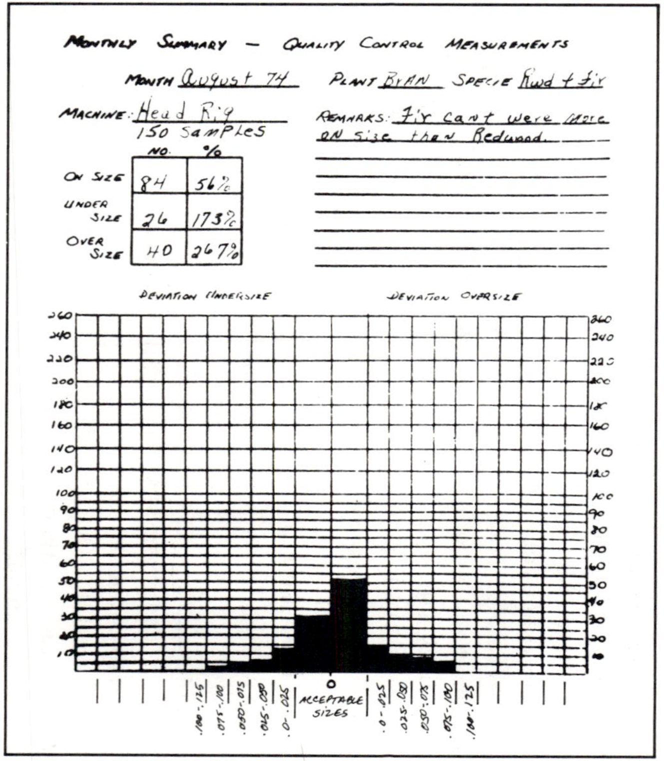

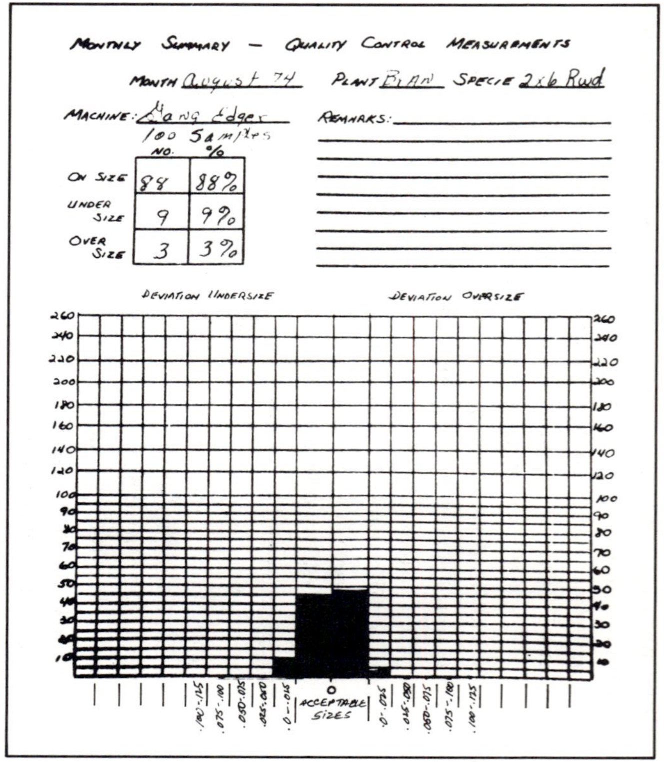

Left: This graph for the headrig shows an unacceptable spread of sizes. *Right:* Concentration on this graph for the gang edger is in the on-size columns. Undersize and oversize fall within acceptable tolerances—plus or minus 0.025.

The people side of quality control

by James L. Russell

At Idaho Forest industries (IFI), we feel a quality control program is only as good as its people. We believe, in fact, that quality control *is* people. To work, a quality control program must be initiated by the people in management. It is introduced to the people in the production areas by those the management team picks for the task. We have tried several different approaches. We have found that machines are important and new machinery helps. But the best machinery you can buy is worthless unless the people operating it know what you expect from them. You need capable people in both management and quality control, as well as in the supervisory positions in production.

The people in the quality control area at IFI are really no more than a monitoring group—we monitor improvements in day-to-day production. A main function is communication. We take the things management expects and present them to the people in the production areas in such a way that they not only can understand them, but can live with them.

We feel selection of people to implement quality control can spell the difference between a successful quality control program and one that really will not accomplish much. We feel this because we tried the old approach of backing into quality control. For a long time our quality control manager was the head grader, but the people in the sawmill did not believe a man from the planer department could teach them anything. And he was not the kind of fellow who really grabbed the bull by the horns and tried to overcome that prejudice. Now we use a different approach. Our present quality control manager is well-versed in every phase of production—from the stump to the boxcar. We think we are finally on the right track after four years of searching.

We feel the position of quality control supervisor should be filled by someone who is familiar with the complete sawmill operation. The most important reason is that the production supervisor will accept advice, comments, and suggestions much more easily from someone who has struggled through the same problems than he will from someone who was picked out of the office. We tried that approach, too. It did not go over well. So our theory on selecting quality control personnel is to pick people who know what they are talking about. Then when they talk to the people in production, the production people will listen—or at least they will not run them out of the mill.

Communicate goals in language mill workers understand

Communication is most important in conveying management's desires to the production supervisors. This must work both ways. It does not do the plant much good for the supervisors to understand what you want if they are not getting it for you. We have a daily monitoring plan by which we communicate with the supervisors. It is very similar to the procedure described by Bob Mardock: When the quality control man goes into the mill, he locates either the maintenance foreman or the supervisor of the particular department. Anything the quality control man intends to report to management is reported first to the person in charge of that particular department. The intent of the program is not to get people in trouble. Everyone is simply trying to come up with a program that will make us some money.

We tried initially to use percentages of recovery to communicate goals. So much lumber is 1 percent of our production and a 1 percent increase in production is worth so many dollars to us. It sounded great to management but it did not go over very well in the mill. One percent of what? What does 100 million mean to a man who is tailing the edger? So now when we introduce a measurement standard to quality control, we use the dollars-and-cents approach.

Of course, simple tolerances are the basis for quality control. But to emphasize how important tolerances are, we actually compare the dollars-and-cents values of a Sterling Idaho and a Standard Idaho board grade. For instance, the difference between a Sterling Idaho 1x12 16 feet long and a Standard Idaho 1x12 16 feet long at current market conditions is about $2. And it takes only a dig with a lift truck or someone dropping the board on the ground and getting it muddy to lose that $2. We find our people accept this dollars-and-cents approach considerably better than they did the percentage approach. This might not be true everywhere, but it is true at Idaho Forest Industries.

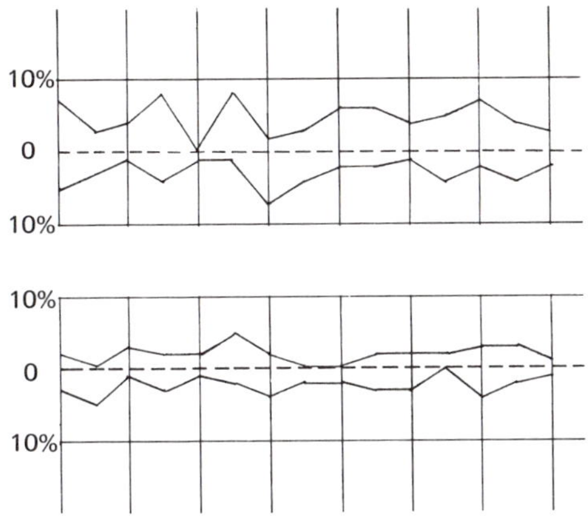

Figure 11.1. Graph used to monitor lumber grader trainee's progress. Upper section shows first three months' accuracy; lower section, second three months. Effort was quite successful.

Also, quality control people must understand the science of quality control—standard deviations, best opening face measurements, things of that nature. We use graphs and charts to train our people.

Figure 11.1 shows a graph we used in training a lumber grader. This graph covers a six-month period. You can see that in this case our effort was very successful. The graph does not always look this good, but in this case we were very happy with it. We keep it posted on the wall in our shipping office because it is exceptional. We use graphs similar to this for measuring progress in many areas—dry kilns; scheduling; grade recovery at the planer; and, to a degree, sawing variations at the headrig and resaw.

When we began our quality control program at IFI, we used terms with which the production people were not familiar. For instance, when you start a program, not many production people will know what you mean by "standard deviation." They might not equate overrun with the lumber recovery factor. These are things quality control is responsible for phasing into the operation, but in a manner that will not alienate the production people.

Guidelines must be set by management

We feel the job functions of the quality control people *must* be spelled out by management in such a way that both management and production have a very clear understanding of what the program is expected to accomplish. The responsibilities of the first line managers should be spelled out also. It is much easier for production people to accept a definite plan from management than to have a quality control man simply move into their department, start taking measurements and making notes, and then walk out with the information without their knowing why or where it is going. We keep our people posted. Communication is the name of the game.

We have laid out a program for each department supervisor indicating where quality control and production balance. In other words, we have set up a standard for each machine. If we find at any time during the day that standard is not being met within a certain percentage (we might set up a maximum 10 percent deviation in some places; others are as little as 1 percent, depending on the machine's importance to the operation), the machine is shut down until it is repaired or taken care of. The same is true with operators. If we find we cannot train an operator to do the kind of work we must have, we train a new operator. It is that simple. At IFI, quality control is at least equal to—and in a considerable number of cases, takes precedence over—production, when it comes to a choice. But we feel that both quality control and maximum production are possible if operating personnel are properly trained, so we place a great deal of emphasis on training.

Training a major factor in quality control

Training the crew is definitely the first line manager's responsibility. He is the one the crew will come to with their questions. They will not go to the foreman's boss. They will go to the foreman. We expect the foreman to train his own crew. We give him all the help he asks for, all the information he needs or at least all the information we have. We provide all the assistance he will accept, but we do not force it on him. We feel the first line manager is probably the most important person in a quality control program. He is more important than the quality control people who come in and monitor. As many supervisors describe them, "All they do is find fault and run to the boss." But that is a part of the job, nevertheless.

At Idaho Forest Industries, we are attempting to carry out a communications program that really works by writing a quality control manual for each machine. By the time we had finished the first one, it had turned into an operator's manual. It is actually a training tool. We did not start with this intent, but this is what we wound up with. The supervisor of the department, along with the quality control manager, prepares the rough draft for the manual. We also write manuals for operating areas where there are no machines—the green chain or the planer chain. Then at our weekly quality control meetings, which are sometimes dinner meetings (our management is behind this 100 percent and is not adverse to spending a few dollars to keep people interested), we go over the rough draft. We review it and/or revise it. Every supervisor concerned (we have more than one mill) is involved in revision or the discussion of whether there should be a revision. When we are satisfied that we have what we want, we give the rough draft to the people who will operate the machines. When they have had a chance to read it, we go back to them and get their suggestions. If their suggestions are useful, we put them in the book.

Figure 11.2 shows two pages of a handbook we wrote for the bucksaw. In a mill like ours, you would think this such a simple operation it would not need a manual. We have a stud mill where everything comes in, supposedly, in 8-foot 2-inch increments and we have a random-length mill where the logs come in all lengths. This handbook was written for the stud mill. Some may think it ridiculous to spend this much time on such a simple operation. But we feel communications are important enough to warrant the time. Having the man know exactly what we want him to do is important to us. No percentages, no dollars-and-cents figures are involved. We just tell the man, "This is the way it should be done." We also tell him why it should be done that way. These handbooks are available for the operator to look at, they are available for the supervisors to use as training tools and as communications tools, and they are also available to the supervisors as a "club" if they have a fellow who is stubborn.

diagram 1

(log must be positioned against the log stop before bucking)

3. Random length logs, logs with broken ends, and logs with stump pull require special consideration. Such logs must be positioned correctly before bucking. (diagram 2&3... random length) (diagram 4&5...broken end) (diagram 6... Stump pull)

diagram 2

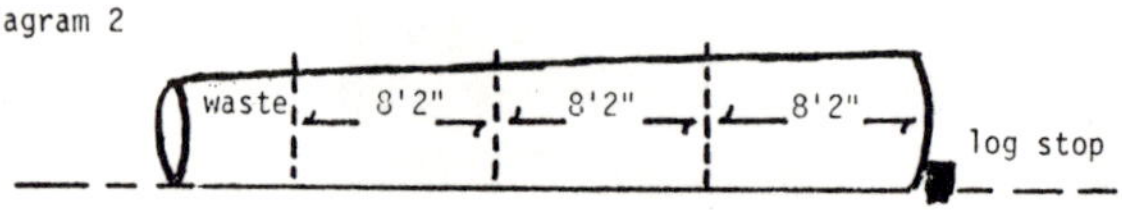

- random length log; large end presented to stop-
8' length should be bucked from large end of log. Waste should be located on the small end of the log.

diagram 3

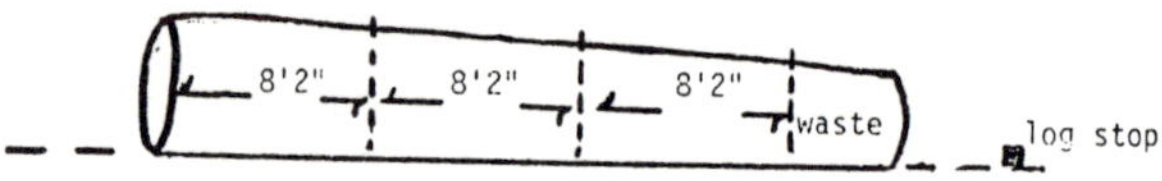

- random length log; top end presented to stop-
Such logs will be marked with a visible color prior to placement upon the log deck. This will enable the buck-saw operator to recognize random length logs and position them in a manner to maximize 8-foot lengths from the larger end of the log. Such logs will be positioned with the aid of a closed-circuit television set. The TV will serve the basic function of a log stop on the back end of the log. Extreme care will be necessary to obtain logs of the proper length. (8'2" - 9'0")

Figure 11.2. Pages from Idaho Forest Industries' bucksaw manual. Handbooks such as this are written for each operator station or area.

diagram 4

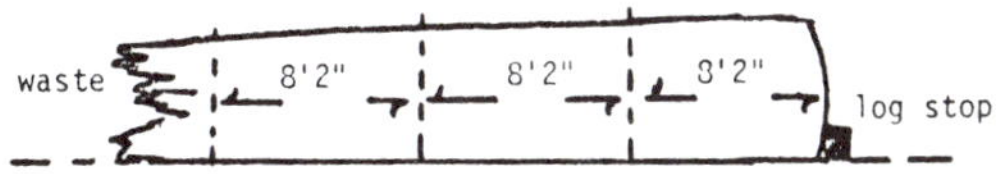

log with broken end; sawn end presented to log stop - 8' length should be bucked from large end of log. Waste should be located on the small end of the log. Treat the same way as a random-length log.

diagram 5

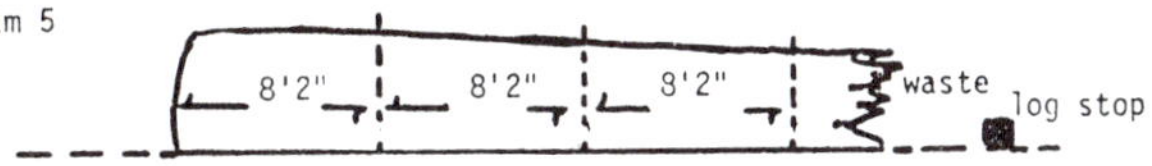

log with broken end; broken end presented to log stop- the log must be positioned in a manner to maximize 8 foot lengths from the larger end of the log. Refer to diagram 3, random lengths.

diagram 6

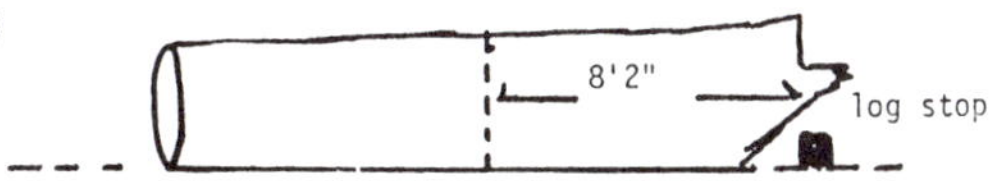

log with stump pull presented to log stop- it is necessary to position such logs by sight. Proper alignment should be as shown in diagram 6

4. Operate waste kicker to remove short and unusable material from the log conveyor. Such material will include pieces shorter than six-feet. Pieces 6 feet and longer should be sent to the debarker.

C. Observe all machinery in buck-saw control center and notify millwright if machinery is not working properly. Following is a maintenance schedule for machinery to be checked in the buck-saw control center.

IFI believes that communication is vital to quality control and considers these manuals an important part of its program.

We feel these manuals are probably the most worthwhile element of our quality control program. They present our quality control guidelines, enumerate the kinds of quality control checks we feel are needed, give instructions on both the method and the frequency of making those checks, and indicate the proper action to be taken when it is discovered that something does not meet the specifications.

When a program of this type is completely installed and totally accepted by all concerned, your quality control program will be working because your people are doing the job for you.

People, machinery, management—three building blocks essential to a quality control program. Now let us see how it functions. Let us put quality control into the perspective of an everyday sawmill operation—from the log deck to the dry kilns and lumber yard.

Quality control from log deck to dry kiln
by Robert W. Nix

A major philosophy of Diamond International Corp. is to produce a *quantity* of *quality* products. This is a contradiction to the adage that a mill can have either production or quality. Diamond, however, is a firm believer in the ability to have both.

I consider Diamond's Northwest Lumber division to be a combination of medium-size operations, producing about 170 million board feet of lumber per year in three random-length mills and a small salvage mill. This moderate size is a plus for practicality in our quality control efforts due to the relatively small number of people involved. A small number of responsible people from top management to line supervisors can easily be made aware of quality control problems and the steps needed to prevent them.

Five basic management areas of our organization are involved in quality control:

Northwest division general manager
Mill management and supervisors

Sales management
Logging management
Division quality control staff

Information on product and process quality flows back and forth among each of these departments. As part of the division quality control staff, I usually deal directly with people in all the above management areas on a day-to-day basis, except for logging management. For a successful quality control program, it is important to have direct access to top management.

I would like to focus on two ideas that Diamond International has implemented in promoting quality control at both ends of the production flow from log deck to lumber yard. One is relatively complex and expensive; the other is quite simple and inexpensive.

Log sprinkling improves recovery

The extent to which Diamond mills utilize water log storage and sprinkled log decks for *all* species is rather unusual in our region. Several mills around Coeur d'Alene Lake, Idaho, use open-water log storage, but few mills sprinkle their log decks as much as we do. I feel that sprinkling logs is most practical for small- to medium-size operations for two reasons:

1. Water supply needed for the volumes required.
2. Handling runoff volumes to meet pollution standards.

Diamond has about 40 acres of open-water log storage and about 50 acres of sprinkled log landings in the Northwest division. Our approach to log storage is similar to that used in Scandinavia. We try to maintain the logs near their natural moisture content to increase grade and volume recovery. Processing in the mill is made easier, also.

For comparison with our operations, some of the photographs included in this section show Scandinavian applications of water log storage and sprinkling.

An example of the adverse effects of not sprinkling was seen at one of our mills. Due to a delayed startup, some unsprinkled logs in inventory were up to two years old. The logs were dry, hard to

Top: Diamond International's log sorting operation at north end of Coeur d'Alene Lake, Idaho. *Bottom:* Log storage bay, behind the sorting operation. Northwest division has about 40 acres of open-water storage.

process, and badly end-checked. As a result, production was down; volume recovery was also down, due to trimbacks to make grade; and sales were difficult because of undesirable length tallies due to the trimbacks. We received more than two dozen complaints from customers on end-check frequency alone, although technically most individual pieces were on-grade. Our sales manager has an entire file of Polaroid pictures showing checks on the ends of loads.

We began sprinkling logs of *all* species at that mill, and now even the older logs from inventory can be processed with relative ease. Recovery of grade and volume has improved considerably and desirable tallies make sales easier.

We feel Diamond International's log yard at Superior, Montana, is a practical application of log sprinkling on 20 acres of landing. Sprinklers are on pilings where they are relatively safe from

Top: Debarking and sorting operation at Soderhamn, Sweden, where logs are barked, sorted by diameter, and stored in bundles in open water. *Bottom:* At Skutskar, Sweden, bundled logs are stored in water and portions above water are sprinkled.

damage. There is less danger of accident because no one has to climb on log decks to install sprinkler pipe. Each row of sprinklers had double shutoff valves so the back half of a deck can be kept under sprinklers while logs are removed from the front half. The cost of this system was about $10,000 in 1971.

We feel quality benefits have more than offset the cost and problems of meeting water pollution control requirements for our

At Diamond's 20-acre Superior, Montana, log landing, sprinklers are installed on pilings to protect them from damage. This also eliminates need for personnel to climb on log decks to install or check sprinklers.

log storage areas. Our lake storage has not had any major pollution problems, and we are taking steps to avoid pollution at our sprinkled landings.

The Superior landing sits on a large sand-and-gravel deposit in a canyon. Water is pumped from two wells. Most runoff simply sinks into the ground. Shallow ditches feed into two settling ponds to handle excessive runoff in the spring.

Our Albeni Falls, Idaho, landing is located along the Pend Oreille River, and is sloped away from the river to prevent runoff from entering the river. The Albeni landing pumps water out of a large (8,000 to 10,000 gallons) sump. Shallow ditches feed sprinkler runoff back into the sump. This is supplemented with river water at intervals to offset evaporation losses. At the request of the U.S. Environmental Protection Agency, a settling pond for excess runoff is to be built at the Albeni landing soon.

I will leave the sawmill area of quality control to others. I would like to say, however, that personnel training and machine accuracy cannot be emphasized enough in the sawmill with today's raw material costs.

Quality control data increase kiln efficiency

At the other end of the production flow, Diamond has implemented a simple and practical quality control procedure in the dry kilns and planers. Whenever we check the moisture content of lumber going into the planer, we also record a sample of the grade recovery behind the planer, the kiln schedule by which the lumber was dried, and the position of the lumber crib in the kiln. Comments are also included on type of defects and any extraordinary circumstances affecting the sample.

We use this information to analyze kiln maintenance, kiln schedules, and lumber-handling practices and to correlate kiln performance and planer grade recovery. The information obtained has been invaluable to quality control and mill management. Since we started this procedure a little over a year ago, we have shortened and adjusted kiln schedules, generally decreased moisture content ranges coming from the kilns, and improved grade. Very rarely is roller split or end-split from drying found in our mills now.

An additional benefit from our efforts in kiln drying quality control has been increased planer efficiency. Our planers have experienced 5 to 15 percent increases in volume throughput. Due to reduced breakups in the machine, we have been able to increase feed speeds on our 20-knife planer about 8 percent.

Quality control is needed in every phase of a mill operation. I believe that application of log sprinkling and kiln quality control procedures would provide maximum benefits with minimum expense for small- to medium-size operations.

The quality control program described next is at work ensuring that standards are met in the world's largest particleboard plant. Let us look at how to pull the whole system together so that quality control is in fact in harmony with production, management, and the banker.

Quality control in harmony with other considerations

by Frank John Banducci, Jr.

I want to give you a general idea of our plant operation and also to explain how we put our quality control program together and how we make it work. Our success with the program comes from experience and much hard work on the part of each individual involved. No one person can take the credit alone. Everyone works together as a team.

Roseburg Lumber Co. is one of the nation's largest independent producers of lumber, plywood, paneling, particleboard, and wood chips for pulp and paper manufacturing. The company has six plywood plants, two sawmills, a stud mill, the industry's largest particleboard plant, a plywood panel finishing plant, a particleboard finishing plant, and (under construction) a low-pressure laminate plant. Roseburg Lumber employs nearly 4,000 people. The particleboard plant alone, when in full production, employs more than 300. The company is one of the largest in Oregon.

We have two particleboard production lines at Roseburg. Line No. 1 is 12 years old and has a 5-foot by 18-foot 24-opening press. It runs 24 hours a day, seven days a week. There is one 8-hour

Mat moving down the line at Roseburg Lumber Co.'s flakeboard plant. This line produces 9.5 million square feet (3/4-inch basis) monthly.

maintenance day a week that we count on for preventive maintenance and repair if needed. Line No. 1 produces 9.5 million square feet per month (3/4-inch basis). We use 3 million pounds of resin and 17,000 bone-dry tons of wood per month on this line.

Line No. 2 is four years old. It has a 20-opening, 8-foot by 24-foot Siempelkamp press and produces 13 million to 16 million square feet (3/4-inch basis) per month. We use 9 million pounds of resin and 22,000 bone-dry tons of wood per month on line No. 2.

Our principal products are Industrial, Underlayment, Countertop, FiberCore, and Commercial grades. We are diversified in that we can produce many different types of boards, 100 sizes, and 11 thicknesses. We need 35 railcars a day to ship our total production.

Quality vital to competitive market position

Quality is the key word at Roseburg Lumber's flakeboard plant. We understand that without a competitive quality product, we cannot survive in today's marketplace, nor could we in the future.

There is no margin for error or for lax or shoddy work by anyone, at any level in our organization. To insure that that philosophy is reality, we have set up a workable quality control program that encompasses the entire organization. No area has been left out. It does no good to have quality control in one or two areas only.

From our raw materials through all production phases (including maintenance) to when the product reaches the customer, quality is stressed continually. The goal of our quality control program is zero defects.

Few products are made, or human functions performed, without the setting of reasonable standards that allow us to work within a given range. Through product development, during actual production runs, and through research, we have set workable specifications and tolerances. Everyone in the organization knows what is fully acceptable and what is not. Precise standards are necessary and all work must come within the set allowances.

Specifications are continually tightened to improve our product. Product improvement will never stop. Every production shift must manufacture the product the same way. We cannot have each person doing his own thing. We must have uniformity.

Operator on No. 1 line is as important to quality control as production foreman or quality control staff. He sees that precise specifications are constantly maintained. Receiver (indicated by arrow) tells operator the weight of mats going down the line.

Although we have several different departments throughout our operation, there is no division. We are all after the same end—a consistent, uniform, quality product at a reasonable price. Production and quality must go hand in hand. You cannot have one without the other.

Management support important factor in success

My boss is Bob Crawford, the flakeboard plant manager. In the 10 years I have worked with Mr. Crawford, he has never waned from wanting a quality product. The same holds true with production. Mr. Crawford wants all the production the plant can produce, but not at the expense of quality. Top management's attitude is the key to any company's success or failure. Mr. Crawford continually stresses the immense job that faces any production facility to all his people, including the laborer on the floor. He personally spends 15 minutes talking with each new employee—and he is a busy man—to tell him what we expect and what type of an organization we have.

This is why a quality control department must be a focal point throughout the entire process: It is the eyes of management. When top management is behind, pushing, and in front, leading, the program will be successful.

Another important key is our plant superintendent, Keith Horner, who is in charge of production. Keith and I work very closely together. We work as a team for quality *and* production and communicate probably more than any two individuals in the plant. We are both concerned for all areas of the plant—never pointing fingers but solving the problems and moving ahead. Any changes made in operating procedures go through both our departments.

Good communications are essential—not just between management and the production foremen, but with the people on the mill floor. You cannot have production without quality nor can you have quality without production. We have teamwork throughout our organization, from supervisor to laborer. We cannot accept anything less than a quality product. Our philosophy is: "We manufacture quality into the product." This requires that we

Although dryer room operator is not member of quality control staff, he and quality control supervisor communicate closely. Any changes in procedure must be made with agreement of plant superintendent.

know from beginning to end exactly what is happening in the production process.

A number of factors are keys to our success with a quality control production program:

1. Top management's attitude.
2. Communications.
3. Everyone's attitude.
4. Organization, each day.
5. Enthusiasm and motivation.
6. Working relationship and training.
7. Responsibility.
8. Accepting corrective criticism.
9. Self-evaluation.
10. Did you have control?
11. Specifications.
12. Cost analysis and control.
13. Statistical data–history.
14. Safety.

There are more than 25 variables that will directly affect the end product and they must be under control.

What does all this boil down to? It means money in the bank, jobs, and a future for all of us.

Where is quality control needed?

Now for a brief description of our operation and the place of quality control in it. Our raw materials are housed in silos. About 45 to 48 mills, including our own, supply us with raw materials. Trucks bring in the material from within a 150-mile radius of the plant. Every truck is weighed and samples are taken from each truckload. We know what we are getting in every truck. One of our storage facilities will hold approximately 30,000 bone-dry tons. Our older units will hold about 10,000 tons. A surplus stockpile can hold 40,000 tons, so our total storage capacity is about 79,000 tons overall.

The trucks unload into truck dump facilities under the charge of one man, who is under my direct supervision. I handle all the

A quality control man is in charge of raw material truck dump facilities. Conveyors and feeding devices can be monitored on this panel. Chart at left keeps track of material in each silo.

raw materials that come into the plant. It is my responsibility to see that they go to their proper silo. We use a chart to keep track of the material in each silo. One man under my supervision watches every truck for safety items, as well as for the condition and type of raw material.

We use magnets to collect some of the metal in the incoming material. It is sometimes hard to believe the amount of foreign material in our raw material. We have a control program in which we take samples, send them to the supplying mill, and say to them, "Will you please help us clean it up." This program has paid off.

After the material has come into the mill, we take a sample of each truckload to the lab. The sample is weighed, put into the oven for 24 hours, taken out, and the data recorded. This information is fed into an IBM machine, producing a readout which we send to the customer. In this way we know what we are getting and the mill knows what it is supplying.

Quality control in our drying room facilities is also under my supervision. My people can communicate with a dryer operator at any time, but quality changes must go through me so specifications can be set for a better product.

Specifications needed for each facet of production

Specifications are set for the size, geometry, and moisture content of the fine material produced for the face of the board. They are precise to within 0.5 percent. The material for the core is larger than that for the face. The flakeboard plant receives about 15 different species with numerous moisture contents that must be controlled from the beginning. We do not wait until the material gets into the main plant to begin control. We have a quality control room in the dryer room. There we make fines checks for moisture content—and the moisture checks must be within plus or minus 0.5 percent.

We record all this information on charts and forms. We do not throw data away; we use it. It serves as history, but even more important, it is used to set up future specifications. Quality control is in charge not only of the raw materials, but also we must understand the machinery—its capabilities and capacities; what we will get out of it; what we can put into it. The information collected on the raw material is checked by me daily. I check this

Sample of all incoming raw material is taken to lab, where it is weighed (*left*), dried for 24 hours (*right*), and the results recorded. Printout tells customer what he has supplied, and is used for mill records.

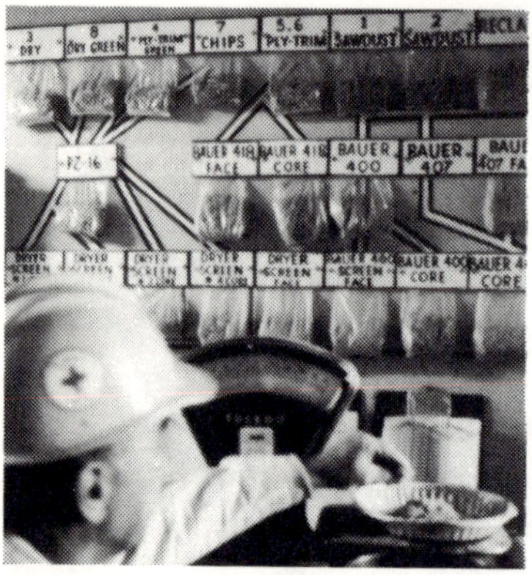

In quality control area in the dryer room, fines are checked for moisture content (*left*) and information is recorded (*right*). Data collected are used to set future specifications. Moisture checks must come within plus or minus 0.5 percent of target.

information about three times a day, and must sign and double-check everything. If something is out of line, I get in touch with the proper people and we sit down and discuss what we can do to correct it or improve it.

Resin mixing is done with probes—in other words, everything is automatic. Even so, we check it personally each shift. This is a quality control check. The resin, water, and wax must be mixed precisely. The mixture is sprayed into the blenders through nozzles. After grinding, the fine material falls into a chute and enters the blenders. There it is mixed with the resin mixture. This is the material that will form a mat on the line.

Floor personnel a part of quality control

The operator on the No. 1 line is as vital to the quality control program as the quality control man and the production foreman. The production foreman has a check card that he must follow. It gives the precise specifications to be maintained. Specifications have been set for every piece of machinery in the plant—there are no exceptions to the rule. A small receiver tells the operator the weight of each mat going by that will be pressed into a board.

The mat moves on down the line (it is called a mat; it is not yet a board) and is trimmed to size. We use magnets to extract foreign

Top: When boards leave press, thickness is checked with calipers by quality control. After sanding, thickness must be within 0.005 inch of target. *Bottom:* Stacking specifications are exacting, because of danger of warp. Boards are checked in this area by quality control, production foreman, millwright, and floor man.

metal from the mat. Even with the 30 or so magnets we use upstream, we still pick up a large amount of metal. Some even gets by the magnets at this point occasionally.

We have another quality control point at the 24-opening press on the No. 1 line. The operator again is as vital to quality control as the quality control man or the production foreman. He has exact specifications to follow for production and manufacturing.

After pressing, the mat is then called a board. A quality control person uses a set of calipers to check the thickness out of the press. This is a very, very important control point. Thickness is important in flakeboard manufacture because you take only so much off with the sander. After sanding, thickness must be within plus or minus 0.005 inch of target.

The board then goes through carbide saws, where it is cut to size. We can cut it into blanks or to any dimension we want. After cutting to size, the boards are stacked. At this point the board is checked by the quality control man, by the millwright, by the foreman, and even by the floor man on the end of the line who is also the jitney driver. They check for all measurements. They check board thickness, uniformity, look, feel, and weight.

We have exact specifications in our hot-stack area because of warpage. First thing each morning I fill out a card on the production in the hot stacks. I walk down the rows, checking for anything I see that may hurt us—color changes, saw marks, anything that may affect the board's reception when it reaches the customer. The board stays in the hot-stack area for 24 hours before it goes to the sander. This gives it a chance to cool down and gives us a better sanding job.

Laboratory tests help maintain control

After the board is cut up, a sample is sent to the lab. We have to meet industry-wide specifications on stiffness. We have quite a bit of testing equipment—we consider our lab sophisticated, but I am sure there are a number of areas in which we could improve. The quality control man takes the sample and with a micrometer and other measurement tools (and his specifications book) goes through the testing procedures for weight, measuring, and moisture content. All our plant standards are higher than the ASTM or government standards. Our standards are higher because we feel this is where we need to be in the industry. Since Mr. Crawford has been plant manager, we have kept one sample from every production run made.

After board is cut up, a sample is sent to the laboratory for testing to assure conformity to specifications. Modulus of rupture (*top*) and internal bond (*bottom left*) are checked using a Tinneus Olsen. A sample of each production run (*bottom right*) is kept by quality control.

Positive feedback important to quality control

As quality control supervisor, I am in my office about 10 to 15 percent of the time. I spend the remainder of my time out in the plant looking and watching. If things are going well, everyone gets a pat on the back. We are not always chewing people out. If something is not going as it should, yes, they will hear about it. But we do it in such a way as to get the point across without making the man feel we are "telling on" him. We are all after one thing: making a competitive, good-quality product at a good price so the men in the plant will have a job 10 or 20 or 30 years from now.

Stan Kuhn, my right arm in the lab, helps me with filing, recording of information, and running the statistical data. We use plenty of charts and forms. We are also in charge of the safety program in the plant. That program is just as important to quality control as anything else in our operation. We also keep a record of grade recovery. Every panel is graded in our plant.

Approximately four years ago, we began something we call the "Flakeboard Plant Manual." Bob Crawford initiated the idea. I did

Strapping standards are precise. Clips must be placed under tie ends to protect boards from damage. Stickers must be in place to prevent warp.

A final quality control check is made on each railcar before shipment to customer. Appearance of product is as important as internal quality.

the bird-dogging, but everyone in the operation helped write this manual. It contains everything essential to the operation—mechanical, electrical, and quality specifications. We are constantly adding to it—and intend to keep updating it.

Control extends through sanding and stacking

Now, let us return to the sander operation, where the board is sanded to a tolerance of plus or minus 0.005 inch. After sanding and grading, it is checked by quality control, the sanderman, and production. If a sander is not performing properly, quality control has the authority to shut it down until the problem is corrected or found. Sometimes it is necessary to run for a while to find the problem. We will play ball, but we will not stand by and watch a quantity of board be produced that we cannot use. The sander on the No. 2 line is an 8-headed Timesavers; the No. 1 sander is a 6-headed Timesavers. A grader checks every panel.

On our No. 2 line we get 120 panels per pressload. Ten pressloads just about fill a boxcar.

After the board is sanded, it can be blanks or 4x8s. It is taken out and stacked in our cut-up area. It can be cut to size. Even unit strapping in the cut-up area has precise standards. We require that small clips be placed under the tie ends. If you do not have them, you will get scratches on the board. Stacking and strapping are done according to standards every time. If the stickers are left out or incorrectly positioned, we will get warpage.

Mel Ray of our shipping department is in charge of both quality and shipping functions in the department. He has charts and forms he must follow, and he checks every car. We have one final check before the boxcar door is closed and the material is sent to the customer. We take this attitude: If we were the customer and we opened that car door, what would we want to see? Properties, internal bond, and so forth, are all important, but it is vital to sales that we produce a good-looking product.

Quality control—do *you* need it? The individuals represented here say *yes*. You need it for profit. You need it for product. You need it for public relations. You need it for recovery. And you need it in the kind of market we have today.

Yes, there is a practical side to quality control. Do you have a program in your mill?

12

Compressed air—without oil-coolant worries

John G. Greenwood, Vice President, Sullair Corp., Essex, Connecticut

The history of the air compresssor is very interesting. The rotary-screw compressor, which without a doubt was a quantum jump in mechanical apparatus for generating compressed air, is an invention and a development of the Swedes. The inventor, Alf Lysholm, was very creative, with perhaps as many mechanical devices credited to his inventiveness as to Leonardo da Vinci. In the early days of the jet aircraft the Swedes made great strides in development of the jet engine. The air compressor and the gas turbine cycle were major factors in that development. Much was done, of course, with centrifugal compressors and eventually with the axial-flow compressors in use today. But Alf Lysholm was seeking á lightweight air compressor (roughly seven ratios or 100 pounds) that would not have the unfortunate characteristic of dynamic compressors—namely, surge or pumping. This occurs at about 70 percent of volume in a centrifugal machine, and somewhat higher in an axial-flow machine. If a compressor is operated in that surge range it does not hold up very well.

Prehistoric man utilized principles of compression

The history of compressors really begins with the dawn of the Industrial Revolution. The piston compressors that survive today differ little from James Watt's steam engine (1765) in principle and in construction. But certainly the first use of compressed air was by developing aborigines who had discovered fire. Some were igniting it by rubbing with a stick; a few were striking fire with a flint; others were actually obtaining fire by compression. They hollowed out a cylinder of wood and a tightly fitting piston which, when hit with a highly propelled club, would reach sufficiently high temperatures from the heat of compression to ignite tinder. This probably took place in the Pleistocene Age and therefore preceded Rudolf Diesel by perhaps 10,000 years.

It is the heat of compression in an air compressor that causes the difficulties. On paper, an air compressor should be one of the simplest pieces of apparatus in a power plant—as simple and as reliable as a water pump in clear, cold water. But it is not. The main problem is that in the compression process, the total horsepower in the compressing of an air-gas mixture goes to sensible heat. A law of Mother Nature reflected in the temperature-rise formulas shows that the discharge of a compressor reaches very high temperatures. In a single-stage piston machine, it is more than 400 degrees Fahrenheit; in a two-stage intercooled machine, it is more than 300 degrees Fahrenheit. These high temperatures are the main bugaboo in compression.

Air compressor types delineated

A piston compressor has a multiplicity of parts, chiefly in the valves. Running at 600 revolutions per minute (and many of them run faster than that now), in a 24-hour day each valve element impacts 1.7 million times. It is a miracle they work at all.

Vane compressors have held a rather important position in the Northwest because of lower hydroelectric power costs. The vane compressor was lower in initial cost—but also lower in efficiency—and has seemed the better buy in the Northwest. There are various types of piston and vane compressors.

Early work on the rotary-screw compressor began more than 30 years ago. There is no doubt about its importance. There are only two working parts in a rotary-screw compresssor: two helical rotors that rotate together. In the dry configuration, these must be separated by timing gears. All the horsepower is in the male rotor; the female rotor is really an idler. In a dry machine, close clearances are required to minimize slip, because there is no mechanical means to seal the high-pressure air from the discharge end back to suction. A dry rotary-screw compressor has no internal cooling and, therefore, has very high compression heat. Compression must be done in two stages with an intercooler, because the close clearances could not stand differential thermal expansion in a single-stage operation.

The next development was the flooded screw compressor. The flooded screw has been used in North America far more than any other type. The dry screw certainly is yielding to the flooded type.

Function of the coolant-lubricant

Yet another quantum jump in air compressor design is development of a coolant-lubricant that is inert, will never oxidize, and does not have to be changed. Prior to development of the new coolant, the flooded screw compressor was internally cooled with mineral oil or synthetic oils. The oil is injected with the inlet air and intimately mixed with the air during compression. Its function is to pick up the heat of compression. Thus discharge temperature, even with a single-stage machine, is normally about 170 degrees. In this new machine internal temperature can be controlled.

Second, the coolant is a seal between the running clearances of the two rotating parts and the cylinder. It is like piston rings in a piston machine. Third, and almost incidentally, it is a lubricant. Lubrication needs within the flooded rotary-screw compressor are marginal, however, because the male rotor does all the work. The female rotor is just along for the ride. The male drives the female without timing gears on a pitchline, as in a gear set, but there is essentially no pressure loading at the theoretical contact point—the pitchline.

This is a zero-wear compressor. The two rotors are the only working parts. They rotate. They have full clearances in the bores and at the endplates. There is no contact—and thus no wear. After many years of operation, it is impossible to see the pitchline where the two rotors theoretically touch. They do not even polish in that area. With properly and conservatively designed bearings to support these rotors, you have essentially a zero-maintenance compressor, with one exception—the dirt formed in the machine due to the temperature rise and the resulting oxidation of mineral oil.

Identifying the source of air compressor dirt

In a screw compressor, perhaps more than in any other type of machinery needing a lubricant or an internal coolant, the compressed air is intimately mixed with the oil, hastening or accelerating its oxidation, even at temperatures of only 170 or 180 degrees Fahrenheit.

Oil is peculiar. It is very much a doctored product. Oil varies from a petroleum-like substance to medicinal mineral oil, which is very pure and clear as water. The list of oil additives to suit it to individual uses is as long as your arm. In your automobile, 18 percent of the oil (by volume) is additives. Additives introduced at refineries vary, depending on the oil source. Additives are hydrocarbons. When one thinks of oil oxidizing, one tends to see it forming an amorphous material—gum or varnish—but many oil additives form gritty particles on oxidation.

If you open a flooded screw compressor cooled with mineral oil (particularly one that has been run hot or one in which bearing filters, strainers, and air-oil separators have not been serviced as required) you will find dirt and grit. The obvious explanation is that the dirt entered with the inlet air through the intake air filter. This is not the case. It has been difficult to dispel this belief in the minds of our own service people—and, in the early days, our own engineers—but it is simply not correct.

With a few scientific calculations, the total amount of dirt that could get by an air intake filter, normally of a 10-micron size, can be determined. It is negligible. The air intake filter, even in dirty industrial areas or around sandblast jobs, really is not a problem. It is easy to separate the gritty particles large enough to cause damage (10 microns and larger) from the intake air. The dirt is generated within the machine.

Many would agree that of all their equipment, certainly of all their so-called power plant equipment, the biggest thorn is the air compressor. It is usually off in a corner; it is ignored; it is an orphan. No one looks at it or takes care of it until it fails or malfunctions. Then a maintenance crew goes to work. In some cases they create more problems than they solve.

So development of a machine with only two working parts aside from the bearings was a quantum jump. The remaining problem with the screw compressor was the dirt generated within the machine by oxidation of the coolant-lubricant. A recent advance has made this a new ballgame.

Research produces new coolant

Sullair is a new company. In 1966, our first year, we shipped compressors worth \$1 million. In 1975, sales were \$50 million. We make nothing but flooded screw compressors, both engine-driven portable units for the construction industry and stationary electric units for other industries. More than 30,000 compressors manufactured by Sullair are in service. This has all happened within the past 10 years. As manufactured by Sullair, the flooded screw compressor is built in sizes from 85 cubic feet per minute to 3,000 cubic feet per minute. Pressures, normally, are so-called 100-pound air, but high- and low-pressure units are also available.

There is no doubt that the piston machine is on the wane, as many predicted 10 years ago. The same is true of the vane machine. There are two types of vane machines—dry and flooded. The flooded vane machine does have the tremendous advantage of internal cooling. Drawbacks, however, include localized hot spots due to frictional wear; the fact that the vane is thrown out by centrifugal force, wiping the cylinder bore; the extreme temperatures reached in the oil film under pressure; and the vanes going in and out of close-clearance slots with friction and compressing air at the bottom of the slot as they push in rapidly. The bulk discharge temperature is about 170 degrees, as in the flooded screw machine. Yet parts of that oil have seen extreme temperatures well above 170. Oxidation in a flooded vane machine is much greater than in a flooded screw compressor.

The new coolant is the result of more than four years of research and development by our company and outside consultants. A host of materials was experimented with and tested. We call the new coolant 24KT. It is as enduring as the name might imply and essentially as pure. It has neither the number nor the type of additives present in mineral oil.

The base material for this new 24KT coolant-lubricant is silicone oil—liquid silicone. Silicone is SiO_2. Its source—the raw material—is sand. It was a curious laboratory phenomenon. The polymers can be made in solid form, semielastomer form (Silly Putty, for example), greases, and liquids. They wind up in all manner of commercial products today. A major manufacturer of silicones is Dow Corning.

Coolant characteristics close to ideal

In liquid form, silicone looks and feels slimy like oil. It was found to have the characteristics that would make it an excellent coolant. At temperatures from minus 60 degrees Fahrenheit to 180 degrees Fahrenheit, its viscosity changes very little. No other liquid has such a flat viscosity curve. Being inert to oxygen, and to most other gases, it would be suitable for difficult, high-temperature bearing applications. It is very difficult for it to emulsify with condensed water. It is about half as flammable as mineral oil. It is nontoxic. It has a very low vapor pressure. If one were to list characteristics for the ideal coolant-lubricant, it would meet all of them—with one exception. It is not a good lubricant.

Dow corning has developed a patented additive—an inorganic lubricant—which makes silicone an excellent lubricant for our screw compressors. If you get some 24KT, however, please do not put it in other types of compressors or in any application where there is a sliding action, particularly if the materials are similar.

Advantages claimed for new lubricant

This means, of course, that a charge of coolant-lubricant that is put into the compressor—provided it is not lost downstream—never needs to be changed.

The dirt in an air compressor cooled with mineral oil is formed within the machine. This shellac-varnishlike sludge plugs the air-oil separator, plugs the strainers, and plugs the bearing filters. The cost of an air-oil separator is not insignificant. The separator must be changed periodically, depending on how your machine is running, the quality of the oil you use, and other factors. If you do not form the shellac-varnish and sludge within the machine, you do not blind the air-oil separator.

Also, after two or three changes of a very tight bearing oil filter, you then change once a year. The differential pressure does not change one wit—we have them running 8,000 hours now.

In a flooded screw compressor, the air-oil separator must handle the flood of oil-coolant that comes off the compressor mixed with the compressed air. (In a 150-horsepower compressor, this is approximately 20 to 25 gallons per minute.) Obviously, it must be separated or you would need a fire hose to keep the sump full of coolant. Trapping that oil requires some expertise in filtration. From the start, our flooded screw compressors were represented to entrain from one-third to one-quarter of the oil entrained by a normally lubricated piston or vane compressor. With 24KT, the entrainment with the standard air-oil separator is one-tenth of what it is with petroleum oil. This is due to other physical characteristics of the fluid—wettability, surface tension, and so forth.

One other point: In applications where completely oil-free compressed air is a must, downstream filtration using simple, long-life separators will remove all aerosols, including hydrocarbons. We have evidence from outside laboratories that there are fewer hydrocarbons in the discharge air downstream from our cleanup equipment than with nonlubricated compressors. That may seem an anomaly, but there are hydrocarbon gases in the environment, even in clean areas. They are certainly present in industrial ambients where compressors are used. Just as the water vapor in this room will condense and form aerosols, so some of the hydrocarbon gases in the inlet to a nonlubricated compressor will, under pressure and after cooling, condense to form liquid hydrocarbon aerosols. Our filter removes them. With a so-called dry machine, they pass through.

DISCUSSION PERIOD

QUESTION: What is the price of the 24KT coolant-lubricant?

GREENWOOD: The price is $43 per gallon, but you get it with the machine and need little, if any, for topping off.

QUESTION: Can 24KT be used in a rotary-screw compressor manufactured by another company?

GREENWOOD: The odds are very good that it would not work. This is not intentional, however. A compressor air end, as we call a compressor unit, has some subtle but very important differences from the normal air end for petroleum oil.

13

Opening the flow from bulk storage containers

Rick Bickers, Sales Engineer, Stephenson Industrial Supply, Inc., Portland, Oregon

My subject is a new product representing a new concept in obtaining flow from bulk storage containers, whether for production or for waste disposal. This product cannot be used in all problem areas, so we must also discuss industrial vibrators. Industrial vibrators have performed poorly over past years in some of your facilities. Although considered necessary evils, vibrators can be useful. But first let us discuss the Big Blaster air cannon.

The Big Blaster is a very simple product with only one moving part, a polyurethane piston, safe and nonsparking. The Big Blaster works off standard compressed air from 60 to 125 pounds per square inch. After it is fully charged, this compressed air can be released in 4 milliseconds, releasing a powerful, controlled energy work force directly into the material. The Blaster works equally well on any product with a density from 5 through 200 pounds per cubic foot.

How the unit operates

Figure 13.1 shows a four-step sectioned view of the Blaster, beginning with the unit at rest. The main control component is a three-way operating valve, normally closed. It can be either manually or electrically operated.

In step 2, we have just opened the operating valve to allow constant air through to the Big Blaster. This air pushes the urethane piston down to the seat. Air is then passed through the 3/16-inch-diameter hole in the piston center into the ASME code-welded pressure vessel. As it fills, pressure on both sides of the piston is equalized and the piston is in suspension.

To release this stored energy, we shift the three-way operating valve to exhaust, removing the air from the area on the control side of (behind) the piston. We have a constant pressure on the tank side, and the piston is blown back. All the compressed air is released directly into the material. Release to atmospheric pressure is completed in 4 milliseconds.

All the air released is used. The initial impact moves material from the air entry location toward the container discharge because the force begins compacting material and continues to compact it until the resistance is overcome. (The Blaster is normally installed in a container pointing toward the discharge.) The resistance will end as the compression of the rathole or bridge is broken. The balance of the unspent air filters up through the mass and loosens compacted areas closer to the bin discharge.

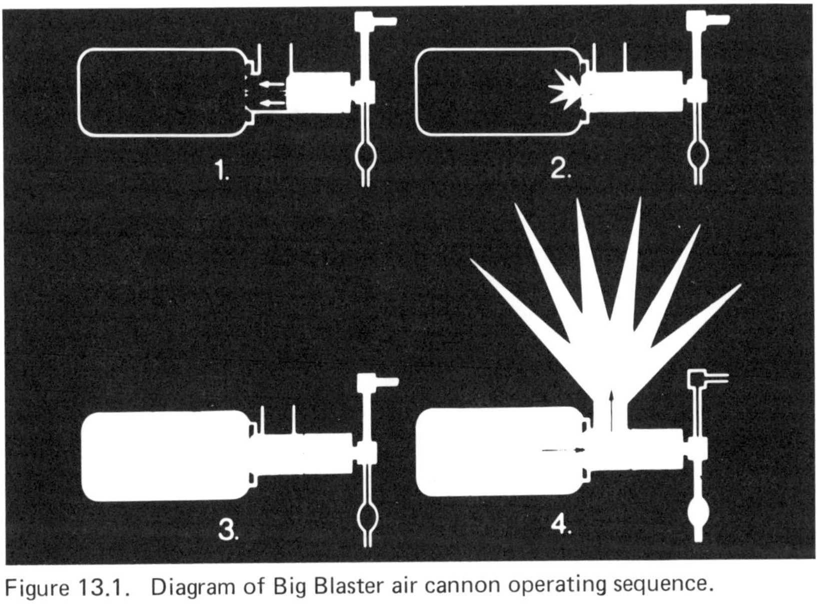

Figure 13.1. Diagram of Big Blaster air cannon operating sequence.

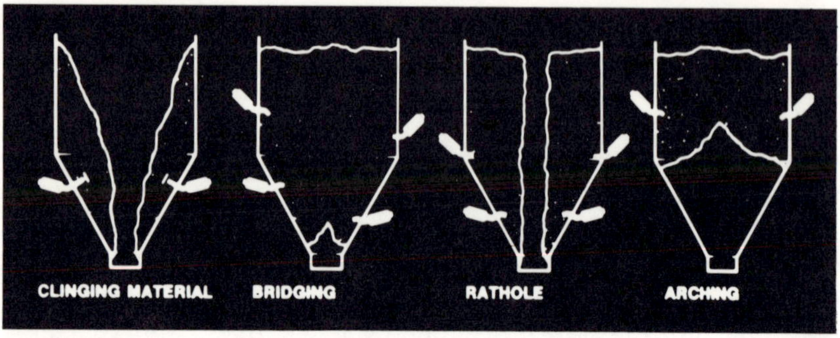

Figure 13.2. Bulk material handling problems in wood products industry.

Identifying the problems

The wood products industry has the most unusual problems in bulk material handling. Figure 13.2 illustrates the problems common in this industry: clinging material, bridging, ratholing, and arching. The figure shows how a full rathole begins. Material starts to cling to the container walls. After time we get a narrow rathole of free-flowing material, with the balance (60 to 80 percent of the storage area) becoming a permanent mass. After a mature rathole develops, you have last-in, first-out flow, with minimal storage.

Finally one day we clean the container by manual means—labor—and sometimes with dynamite. After it is cleaned and refilled, we have 100 percent live storage. We have it, for a week or a month—and then the same problem we just spent money and time to correct reappears.

Sometimes a mass of rathole wall will fall and plug the discharge, giving us a bridge. A further difficulty is that this bridge can be directly at the discharge point, far above the juncture of the slope and the vertical wall, or anywhere between those points. The costs when flow is not available at all are staggering.

So the air cannon can be an invaluable asset to this industry. Its simplicity means practically trouble-free operation, and that is vital. We certainly do not need another necessary evil in the mill.

Dependability a plus

I have witnessed teardown of units cycled 350,000 times, and they show hardly any wear. These units had been operated under controlled conditions, I will grant, but we have units working in the field now for five or six years. Temperature variations from 0 to 120 degrees Fahrenheit do not affect the urethane piston. Moisture does not hamper the unit's function, but too much air-line moisture will reduce the tank capacity and its effectiveness by reducing air space.

Air usage varies from 2.3 cubic feet through 87 cubic feet per firing, based on 90 pounds-per-square-inch compression. As manufactured by Martin Engineering Co., the product comes in six standard tank sizes and three standard discharge sizes–2, 4, and 6 inches. Only the 2- and 4-inch sizes relate to wood products.

Using an 8-inch by 12-inch air tank for the 2-inch discharge, the tank will contain 2.3 cubic feet of compressed air. The 4-inch discharge may utilize tanks of 12 inches by 28 inches (containing 11.4 cubic feet of air) or 20 inches by 30 inches (containing 37 cubic feet of air). The 2-inch size might be used at conveyor transfer points, small hoppers, or cyclones. Tank size for the 4-inch discharge varies, depending on material density and distance from the discharge.

Type of material is a further consideration in proper sizing. A fairly granular substance, like pure sawdust, would lock itself together fairly well, leaving few air voids. For such materials, a tank size of 12 inches by 28 inches would suffice. Materials such as shavings, hog bark, and chips–where there are many voids between the pieces–require a larger air supply, first to fill the voids and then to have something left to do the work.

The cycle requirements on a given container might be two to four times per hour, automated. Based on four cycles per hour, 16 hours per day, 6 days per week, and 52 weeks per year, a given unit would cycle 19,968 times annually–far from 350,000 cycles.

Because of the variety of locations where problems develop, it is often necessary to put in more than one unit. A system, properly located and controlled, must then be used. When this is necessary, the system must be designed for the individual container. Insufficient information is currently available for the mill to develop its own Blaster system.

This product has been marketed successfully for five or six years in other industries, but the wood products industry has been ignored. Yet its need is surely one of the greatest.

Comparison with alternate methods

To what can we compare this Big Blaster? It is reusable dynamite. This industry uses several types of products, rather unsuccessfully, to solve problems for which the Big Blaster was developed.

Foremost are massive vibrators, electric or pneumatic. The electrics require heavy maintenance and are destructive to themselves and to the container to which they are attached. They normally operate at close to 110 to 125 dBA, well above the OSHA standard of 90. The pneumatics are equally poor in terms of maintenance, container damage, and dBA rating. They have one additional fault—they use tremendous amounts of air. Air is never overabundant, and these units are terribly inefficient on the basis of consumption to effective working force produced. Too often they not only fail to remove a bridge, they pack it in more tightly.

Another item sometimes employed is a live-bottom container, a vibrating rubber-mounted cone. This product requires a capital purchase, and all too frequently does not work.

Yet another method is a high-pressure air system. When I say "high pressure," I mean from 500 to 10,000 pounds per square inch produced by a three- to ten-stage compressor. The initial expenditure is again quite large, and a compressor of this type requires expensive maintenance in time and dollars.

And, of course, you can rely on a man with a 16-pound sledge hammer or a poker rod. The hammer is cheap, but using a man for this purpose is expensive—and the results, in terms of damage to the container, can be really expensive.

We cannot sit and wait for a container manufacturer to develop a new design which eliminates these problems. (And in our industry they are compounded by changes in weather conditions and varying moisture contents.) The industry needs an answer now. It should be simple in design and operation, quiet to comply with OSHA, consume little energy but produce much work force, and create no maintenance need in the form of metal fatigue.

I think I have just described the Big Blaster air cannon.

Coping with a necessary evil

I want now to discuss the industrial vibrator, a necessary evil. In some places only a vibrator can do the job, so it is necessary. The evil occurs when one or more of the following operations is *improperly* done: application, motive force, sizing, mounting, or control.

A vibrator has its place on some job requirements, but for each job there are eight or ten different types of vibrators—from a line of several thousand produced by five or six manufacturers—which will give you the desired results. There are also 200 to 300 other choices which will give you only trouble.

Compared with most items purchased in a mill, the cost of an industrial vibrator is negligible, but it can get expensive later.

Vibrators have been in operation for many years. If you require one for an application with which you have no direct experience, however, get help regarding proper motive force, force output, sizing, location, method of control, and mounting. Normally it is not possible to open a catalog, put your finger on one item, say "that's what I need," and get economy and efficiency.

With few exceptions, vibrators should be controlled by automation provided in some way by controls on the item on which they are installed. Vibration at the wrong time can be more harmful than no vibration. A vibrator permitted to run continuously may, for example, break up a low bridge. It may also simply jam the material in more tightly. If it does not break the bridge, it has created a situation worse than the one it is being used to correct. If it does break the bridge, the material will fall from the discharge point, possibly onto a screw area. If the screws cannot remove the material with sufficient speed, all the vibrator has done is to compound the bridge problem.

If this advice can be accepted, you may take the "necessary" for granted and eliminate the "evil" and the problems associated with vibrators.

DISCUSSION PERIOD

QUESTION: Considering the high speed at which the piston operates, what keeps it from beating itself to death?

BICKERS: The piston is polyurethane, which is a flexible material. It comes back against a flat surface, and it is a flat surface itself. The units have been factory-tested at 350,000 cycles at a speed much greater than would be used in an actual installation. If you touch a unit in operation at those speeds, you will feel heat, caused of course by the drag of the piston going back on the wall. But friction does not seem to affect the piston.

14

Dry kiln design and woodwaste recovery

Jack L. Bachrich, President, Intercan Mill &
Kiln Co. Ltd., North Vancouver, British Columbia

Industry has recently been faced with an unheard-of rise in fuel cost. We should examine the consequences with regard to dry kiln operation, at least in the Pacific Northwest. We should consider:

1. Cost of fuels on the West Coast.
2. Heat required for dry kiln operation.
3. Cost of Btus for the kiln in terms of fossil fuels.
4. Utilization of woodwaste instead of purchased fuels for the medium-size sawmill.

I will concern myself almost entirely with the medium-size sawmill, because I feel this group has the greatest present need.

Table 14.1 shows costs of fuels in million Btus. Natural gas goes from about \$1 in British Columbia to about \$2 in most parts of Oregon. I have used round figures in the table and the calculations to follow. I have also tried to put everything on the same denominator, meaning millions of Btus. There is a very solid reason for this. I recall the case of an undoubtedly slick salesman calling on one of my clients and saying, "Why do you buy natural gas, sir? It is costing \$0.50" (meaning per million Btus) "and we can sell you our product for \$0.20" (a gallon, of course—and the gallon has nowhere near a billion Btus).

Table 14.1. Cost of Fuels in Million Btus

	VANCOUVER	NORTHERN B.C.	OREGON
Natural Gas	About $1.00 Little Used	About $1.00 Going to $2.00	About $2.00 Firm
Butane	Not used	About $2.80 '76 Going to $3.00 '77	About $3.00
Bunker 'C'	$2.20	Not used	About $2.14
Dry Shavings			$0.36 (At $6.50/Ton)

In Oregon butane comes to about $3 and is about $2.80 going to $3 in northern British Columbia. Butane is used more in the North because it is the cheaper fuel, as compared with propane. Propane is even more expensive. Butane is more trouble with regard to preheating. It is delivered as a liquid petroleum. It must be heated to form a vapor, and the vapor is subsequently burned. There is apparatus available to burn liquid butane, but the average dry kiln furnace requires it in vapor form.

Bunker C oil has been used in boilers for many years, but even that has had an unheard-of increase in price. I have assumed shavings to be available for about $10 per ton in Oregon; less freight, they come to about $6.50 per ton. They have, in many areas of British Columbia, no value at all. Shavings sell for about $4.50 a ton in northern Quebec. It depends whether you are close to a flakeboard plant. In fact, in many areas shavings, which are such a valuable fuel, have a negative value because obviously it costs money to destroy them in the conventional beehive or teepee burner.

Table 14.2 shows how I arrived at the figures in Table 14.1. I have brought everything down to the same value in millions of Btus. This is quite a simple calculation. If an Imperial gallon of butane has 125,000 Btus per gallon, for example, then 8 gallons are required to produce 1 million Btus. Table 14.2, then, was the basis for multiplication to arrive at the values in Table 14.1.

Table 14.2. Fuel Needed to Produce 1 Million Btus

HEAT	VOLUME
1 - Therm = 100,000 BTU	1 - U.S. Gal = .8 Imp.Gal
10- Therms = 1,000,000 BTU	1 - Imp.Gal=Vol of 10# Water

FUELS	
NAT.GAS	- 1 Ft^3 = 1000 BTU. 1000 Ft^3 = 10^6 BTU
BUTANE	1 - IMP.GAL - 125.000 BTU. 8 Gal = 10^6 BTU
	1 - U.S. GAL- 102.000 BTU. 10 Gal = 10^6 BTU
BUNKER 'C'	1 - IMP.GAL - 175.000 BTU. 5.714 Gal = 10^6 BTU
	Usually use 6.0 Gal = 10^6 BTU
	1 - U.S. GAL- 140.000 BTU. 7.143 Gal = 10^6 BTU
WOODWASTE	Dry Shavings use 9000 BTU/#
	1 - Ton = 2000#. 2000 x 9000 = 18 x 10^6 BTU
	HENCE: 111#= 10^6 BTU.

Woodwaste is an important fuel to our industry. The value of dry shavings has been assumed at 9,000 Btus per pound in Table 14.2. Dry cedar shavings or dry Douglas fir may have a slightly higher value, probably nearer 10,000; other species might have slightly lower values. But 9,000 Btus per pound is a good average. A ton (2,000 pounds) of shavings therefore has 18 million Btus of heating value. Therefore, I arrived at a cost of $0.36 per million Btus. In other words, 111 pounds of shavings (and that is very little) would give you 1 million Btus.

This will lead us then to an interesting conclusion, shown in Table 14.3. The table shows the relative value of shavings if you had to pay for them in terms of natural gas, butane, and so forth. If natural gas is worth $1 per million Btus, then shavings should be worth $18 per ton. If natural gas is worth $2 per million, then shavings should be worth $36 per ton–going to $54 per ton when compared with butane. It would be enjoyable, of course, if you could collect $54 per ton for your dry shavings, but I have not heard that price being offered recently on the market. I hope this table impresses on you the value of the material you are burning or trying to get rid of in these times of high fuel costs.

Cost of fuel to operate gas/butane-heated kilns

Considerable experience is available regarding the cost of operating a kiln on natural gas or butane, especially in northern British Columbia, where this method is quite common. In a lecture to the B.C. Lumber Manufacturers Assn. in October 1965 I stated that 1.25 million Btus of natural gas were used to dry 1,000 board feet of white spruce to 19 percent moisture content. Since then, the new voluntary lumber standard has come about and we should aim at an average moisture content of 15 percent and a maximum of 19 percent. Consequently, additional heat will be required. For the purpose of my calculations, I have assumed an *average* consumption of 1.5 million Btus per thousand board feet of white spruce.

With older stands of spruce, you may use only 1.4 or even 1.3 million Btus. If your lumber is wetter, such as hemlock in the Prince Rupert area of British Columbia, you might be using 1.8 million Btus per thousand. These figures are based on actual measurements. For instance, the use of 1.7 to 1.8 million Btus per thousand board feet of hemlock dried has been measured in a gas-heated kiln. The measurement is very simple. Meter readings are taken and converted to million Btus per thousand board feet.

In all other calculations, I have assumed a kiln-holding capacity of about 200,000 board feet. It is realized that if you take the

Table 14.3. Value of Dry Shavings in Terms of Other Fuels

VALUE OF DRY SHAVINGS - 9000 BTU/# = 18 x 10^6 BTU/Ton

NAT.GAS	\$1.00 / 10^6 BTU	-	\$18.00 / Ton
NAT.GAS	\$2.00 / 10^6 BTU	-	\$36.00 / Ton
Butane	\$3.00 / 10^6 BTU	-	\$54.00 / Ton
BUNKER'C'	\$2.20 / 10^6 BTU	-	\$39.60 / Ton

average 104-foot dry kiln and fill it well with studs, the holding capacity might be a little more–236,000 board feet. On the other hand, if you have mixed dimension loads and gaps in the load, obviously the kiln will contain somewhat less lumber. But we must assume some average, and 200,000 board feet has been assumed. Use your own conversion factors if your fuel value or your loading facility differs from those assumed here.

In Table 14.4, we have assumed a dry kiln of 200,000 board feet, consuming about 1.5 million Btus per thousand. We have also assumed a fairly slow operation–10 cycles per month, or 120 cycles per year. The rest is multiplication.

If, for instance, you had one dry kiln and natural gas cost $1, fuel costs would be about $36,000 per year. By logic, if you had four kilns, fuel costs would jump to $144,000 per year. Now,

Table 14.4. Annual Fuel Costs for Low-Temperature Kiln

EACH KILN 200,000 BD. FT. 1.5 - MILLION BTU / 1000 BM

COST OF OPERATION ON 10 CYCLES PER MONTH = 120 CYCLES PER YEAR

KILNS	PRODUCTION PER MONTH	PRODUCTION PER YEAR	COST OF FUEL PER 1,000,000 BTU	FUEL PER 1000 BM MILLION BTU	FUEL COST PER YEAR
1	2000 M	24,000 M	Nat Gas $1.00	1.5	$36,000
2	4000 M	48,000 M	$1.00	1.5	$72,000
4	8000 M	96,000 M	$1.00	1.5	$144,000
1	2000 M	24,000 M	$2.00	1.5	$72,000
2	4000 M	48,000 M	$2.00	1.5	$ 144,000
4	8000 M	96,000 M	$2.00	1.5	$288,000
1	2000 M	24,000 M	Butane $3.00	1.5	$108,000
2	4000 M	48,000 M	$3.00	1.5	$216,000
4	8000 M	96,000 M	$3.00	1.5	$432,000

Table 14.5. Annual Fuel costs for High-Temperature Kiln

EACH KILN 200,000 BD. FT. CAPACITY USING 1.5 MILLION BTU / 1000 BD FT.

KILNS	PRODUCTION PER MONTH	PRODUCTION PER YEAR	COST OF FUEL 1,000,000 BTU	FUEL PER 1000 BM MILLION BTU	FUEL COST PER YEAR
1	3000 M	36,000 M	Nat.Gas $1.00	1.5	$ 54,000
2	6000 M	72,000	$1.00	1.5	$108,000
1	3000 M	36,000 M	Nat.Gas $2.00	1.5	$108,000
2	6000 M	72,000 M	$2.00	1.5	$216,000
1	3000 M	36,000	Butane $3.00	1.5	$162,000
2	6000 M	72,000	$3.00	1.5	$324,000

double the cost of natural gas (in the Portland, Oregon, area) and fuel cost would be in the neighborhood of $288,000 per annum. If you use butane or propane at $3 per thousand, the figures become horrendous–$432,000 for four kilns. I do not think anyone has more than four dry kilns running on purchased fossil fuel; it is too expensive. But these are the sad facts of life. Even with a medium-production kiln, fuel cost is high.

Table 14.5 shows costs for high-temperature drying and a shorter drying cycle. We have assumed only 48 hours, which is perfectly feasible for northern spruce. As the table shows, there is no pro rata drop in fuel cost, because of faster drying. You might save a little, but not much. Again, for a 200,000 board foot kiln using about 1.5 million Btus per thousand–and I have limited myself here to two dry kilns, because otherwise the figures begin to look too shocking–let us pick the worst example. If you have two dry kilns and pay $3 per million Btus for butane, fuel costs are as much as $324,000 per year.

It seems obvious that with production costs at this level, woodwaste fuel must be considered despite the higher capital cost.

Determining kiln heat requirements

The next point will be to determine the heat requirements of a steam- or hot water-heated kiln. Again we should consider the needs of a typical kiln with a capacity of about 200,000 board feet per charge.

Requirements (in pounds of steam per hour) vary widely, depending on a number of factors:

1. Outside temperature.
2. Temperature of the lumber (frozen lumber).
3. Moisture content of the wood.
4. Insulation of the building.
5. Dry schedule to be followed (pre-steaming, if it is required).

I quote Mr. Bateson of the British Forest Products Laboratory, writing (in 1938) about steam consumption:

> A kiln of the size shown will not normally consume more than 200 pounds of steam per hour. If rapid warming up is required or it is desired to maintain saturated conditions at a temperature of about 212°F . . . a higher consumption (up to 1,000 pounds per hour) *must* be allowed for.

Here you have the problem. The allowable steam load by Mr. Bateson is then one in five—that is, from 200 to 1,000 pounds per hour. The kiln in question had a capacity of 700 cubic feet of 1-inch wood to 1,000 cubic feet of 2-inch wood. Using the higher values, we have 1,000 pounds per hour for a kiln which contains up to 12,000 board feet of lumber, or about 83 pounds of steam per thousand board feet. This is a very high figure indeed.

We must reluctantly reject the high figure of 83 pounds of steam per thousand board feet of lumber. By the same token we must surely reject the low figure of about 16.6 pounds of steam per thousand board feet. In terms of horsepower, this would have been about 0.5 boiler horsepower per thousand for the lower figure, to a maximum of 2.4 boiler horsepower per thousand for the higher figure.

It is easy to say that both figures are wrong and that the truth lies somewhere in the middle. Surely it does, but where exactly is the truth? This is what we want to determine. If we do not know the steam consumption of a kiln, how can we calculate boiler capacity?

I have assumed in this regard 34.5 pounds of steam per so-called nominal steam horsepower. This has nothing whatsoever to do with electrical horsepower. It is an ancient figure which has come down to us in history, but boilermakers still use it. They say 34.5 pounds of steam equal 1 steam horsepower.

The steam requirement which has been assumed in the Northwest for some time is that approximately 1 horsepower might be required for a capacity of 1,000 board feet. Unfortunately, this is not as correct as one might think. For one dry kiln with a 200,000-board-foot capacity, it would come to only 6,900 pounds of steam–let us say 7,000 pounds–per hour. This figure is not too far off if you are working with 15-psi (pounds-per-square-inch) steam and if you are working on a medium- to low-temperature, three-day drying cycle. (In other words, if you are using 15-psi steam, drying northern spruce, and taking 72 hours, this figure is basically correct.) If you are working, however, on high-pressure steam and a shortened drying cycle, you will require additional steam.

Let us work that out for two kilns, which is the more usual installation for boilers using wastewood. Your steam requirement would then be 13,800–say 14,000–pounds per hour, which is somewhat high. This is because with two or more kilns, one kiln will be steaming up while the other one is "on control." The kiln instruments–the steam valves, and so forth–really only work when the kiln is "on control." During heat-up, the so-called control valves are not working at all. They are wide open and permit all the steam available from the line to go into the system during heat-up. You only begin to save steam when the kiln is "on control." Then, hopefully, you use less steam than during the initial heating-up period.

I would like to draw your attention to another method for determining kiln steam requirements–and that is, steam in terms of pounds of water evaporated. I refer particularly to the work of E. M. Wengert of the U.S. Forest Products Laboratory, Madison,

Wisconsin. A German worker named Andersson came to a similar conclusion. *The conclusion:* that the average dry kiln efficiency is only about 0.5. This means that as a basic calculation, to evaporate 1 pound of water out of wood, you must have 2 pounds of steam. This figure, again, is not always correct. Sometimes it is a little on the low side. But do not think you can get anywhere near 100 percent efficiency from a dry kiln.

Why is this? There are a number of reasons:

1. You must heat the kiln building.
2. You must heat the lumber pile.
3. You must heat the machinery and any matter whatsoever inside the dry kiln.
4. You must heat the kiln air.
5. You must heat the fresh air inhaled into the kiln for purposes of humidity control.
6. You have a heat loss arising from the wet air exhausted from the kiln to atmosphere.
7. You have a heat loss when the hot lumber pile is removed from the kiln after drying. That heat has been usefully employed; it is in the lumber. But it is heat lost to the dry kiln installation proper.

Consequently, do not think you can ever arrive at 100 percent efficiency in a drying machine, and certainly not in a lumber kiln. It is simply not possible. If you can achieve 0.5 efficiency—which in round figures means 2 pounds of steam for 1 pound of water evaporated—you are doing pretty well.

The efficiency of gas-heated kilns is probably less than that of steam-heated kilns, as products of combustion constantly enter the kiln. Thus kiln air *must* be bled off, much to the detriment of the level of humidity within the kiln. The same remarks apply to all direct-fired hot-air kilns—that is, all kilns where no heat exchanger is employed.

Operator costs—an important consideration

Having tried to explain how we use heat in the kiln and why we need heat in the kiln, let me make one further observation. In

Canada one type of steam (15 psi or less) is called low-pressure steam; 150-psi steam, which is the usual supply pressure, is called high-pressure steam. The legislation differentiates between the two in terms of the type of boiler operator required. The laws vary from province to province. I believe in the United States they vary from one state to another. In Canada, the law is most severe with regard to high-pressure plants–you *must* have operators. There are a few exceptions for 15-psi boilers in Quebec and Ontario which would suit a mill with only one dry kiln. In British Columbia, however, if you have steam, you must have operators. That is the law.

For hot water, the laws are vastly different. Perhaps pressure has been applied toward leniency in view of the great number of apartment blocks historically heated by hot water. Consequently–and illogically–hot water at 250 degrees Fahrenheit and 50 pounds pressure does *not* require an operator, providing it is running automatically.

The same thing has been translated into the dry kiln field. In Canada, for any reasonable-size hot-water boiler–providing the pressure is no greater than 50 psi and providing the temperature is no greater than 250 degrees Fahrenheit–you do not need an operator. Financially, this will make a big difference, because a supply of boiler engineers will cost approximately $60,000 per year with present wages–I am thinking of a continuous-shift basis. Hot-water operators certainly cost something. You will need some sort of night watchman and the kiln operator might look after the boiler during the day, but the cost of this will be considerably less than the cost of continuous-shift boiler engineers.

Conclusion on steam consumption

Boiler Capacity is based on 2 pounds of steam for each pound of water removed from the lumber. This will result in somewhat lower capacities than might be desirable where only one kiln is concerned. You should have a little reserve in your boiler installation. On the basis of 1 horsepower per thousand, it will work out a little high for two, three, or four kilns. But that is no disaster. A bigger boiler plant is hardly any trouble. The fuel is cheap, as we have established, and if a little more capacity should be available,

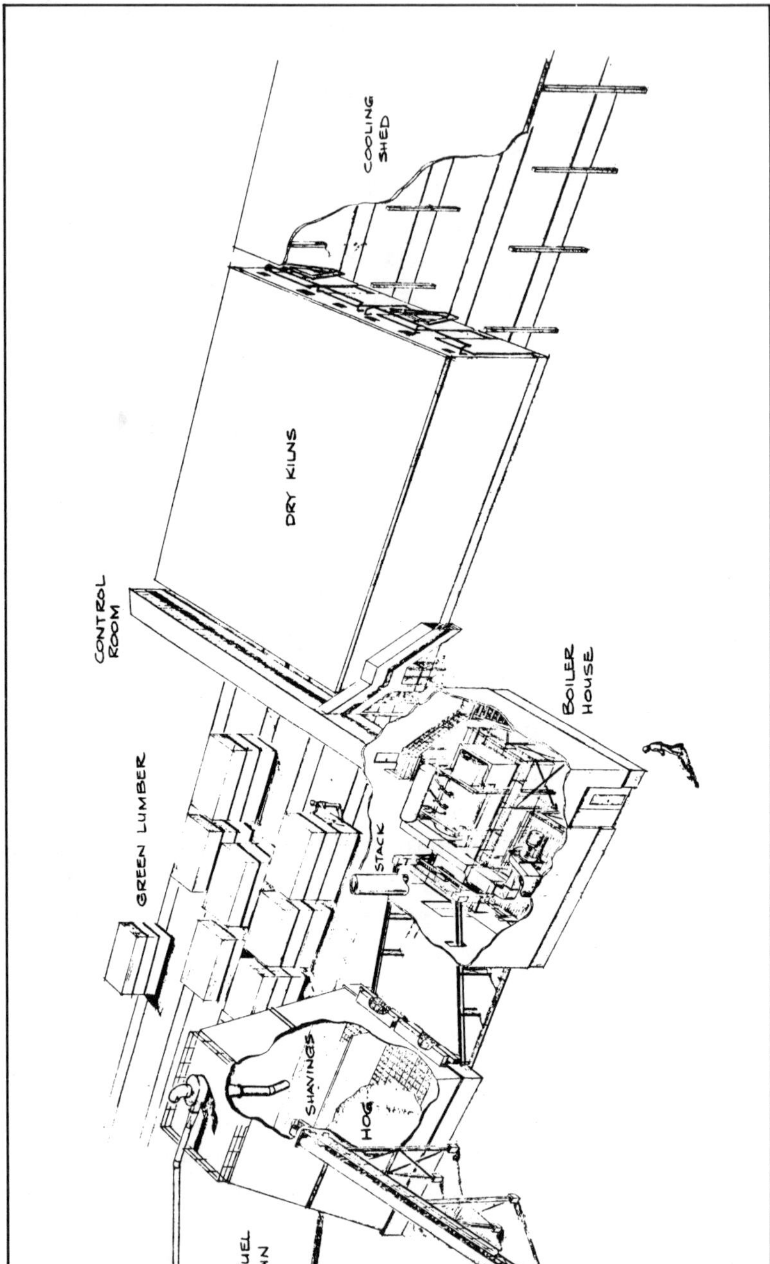

Figure 14.1. Typical dual-fuel woodwaste-fired boiler and lumber dry kiln installation.

your equipment might last considerably longer.

To arrive at some average values, however, a boiler plant of about 12,500 pounds evaporation will in all likelihood be about right for two dry kilns. By all means double it for four dry kilns—let us say 25,000 pounds evaporation for four kilns. The more kilns you have and the longer the drying period (for example, Douglas fir clears), the less the steam consumption. I cannot possibly go into all steam consumption combinations here. But if, for instance, you have four kilns and every fourth day one kiln is emptied, recharged, and reheated, the average steam consumption will be less than if you are operating only one kiln. In other words, you have avoided the peak period. You would not heat up four dry kilns simultaneously; that is simply bad housekeeping.

Greater steam consumption will be needed for high-temperature drying. Again, with a number of kilns, the peak period can be spread over a number of days.

I feel more attention should be given to the woodwaste-burning plant in the above heat range. Using a value of 1,000 Btus equaling 1 pound of steam (approximately), we are thus discussing 12.5 to 25 million Btus per hour.

A plant would either be for 150-psi steam, if engineers are available, or for 50-psi steam and 250-degree Fahrenheit hot water, if engineers are not to be used. There is no sense having 15-psi steam in Canada, as engineers would be required for this size plant. This may differ in other jurisdictions.

The advantage of hot water is the absence of boiler engineers. The disadvantage is a longer drying time, compared with 150-psi steam.

Figure 14.1 illustrates a wastewood-fired boiler installation suitable for from two to four dry kilns. It utilizes a live-bottom fuel bin. Two fuel conveyors are shown: one carrying waste fuel, such as hog fuel; the other, shavings. The fuel then enters the boiler. Instrumentation can easily be installed on the boiler to sense temperature in the furnace, steam demand, and steam pressure. The instruments can control increases or decreases in wet or dry fuel to maintain furnace temperature and to meet steam demand. A pilot flame is provided as a standby. Steam or hot water is then delivered to the kilns through convenient overhead piping to the control room.

Woodwaste requirements calculated

I have shown that we can assume 9,000 Btus per pound of dry shavings. Woodwaste requirement calculations appear in Figure 14.2. You must allow obviously for two days' (48 hours) fuel bin capacity under all circumstances, because of the recurring weekend. You should probably allow for a 72-hour fuel bin capacity to give yourself leeway on the long weekends, about one per month.

The calculations show that a certain size fuel bin will be required, depending on the number of kilns and the woodwaste moisture content. For two kilns, the fuel bin will have a capacity of at least 15,000 cubic feet; for four kilns, double that capacity.

- For dry shavings, assume 9,000 Btu/lb.
- Bark or sawdust will have the same Btu value as shavings, if bone dry. If wet, the net Btu value must be calculated according to the green moisture content of the fuel used. Approximately 1,200 Btu are required to evaporate 1 lb of water. This amount must be deducted from the dry Btu value before a net Btu/lb value can be established.

See also Bramhall, G. 1971. Some developments in smokeless burning. Vancouver, B.C.: Western Forest Products Laboratory.

Swedish sources (Fromo Burners) assume 5 lbs of bark = 1 lb of crude oil, giving the average Swedish bark a value of 4,000 Btu/lb.

- For two kilns assume a boiler with a steam production of 12,500 lbs/hr = 12 million Btu/hr approximately. Assume a boiler efficiency of 0.75. Hence:

$$\frac{12,500}{0.75} = 17 \times 10^6 \text{ Btu/hr}$$

1 ton shavings = 18×10^6 Btu/hr (Table 14.3)

Assume 1 ton of shavings per hr for two kilns

If loose weight of shavings = 5 lbs/ft^3 then $\frac{2000}{5} = 400$ ft^3/hr

If compacted weight of shavings = 10 lbs/ft^3 then $\frac{2000}{10} = 200$ ft^3/hr

For a 72-hr weekend, fuel required is:

72 x 400 = 28,800 ft^3 or 30,000 ft^3 of loose shavings
72 x 200 = 14,400 ft^3 or 15,000 ft^3 of compacted shavings

- For four kilns and 25,000 lbs/hr steam, double the quantities above: 30,000 ft^3 or 60,000 ft^3 would be required. Running light or adding fuel on a 72-hr weekend would permit use of a 45,000 ft^3 bin capacity.

Figure 14.2. Calculating woodwaste required for two or four kilns.

If the bin size required becomes too large—30,000 or 60,000 cubic feet—you may consider running light on long weekends only or bringing additional fuel to the fuel bin during the last leg of the 72 hours. This might involve a separate fuel pile and a front-end loader to add fuel to the bin. I have seen a bin designed for two dry kilns (the bin contained about 30,000 to 36,000 cubic feet of fuel) forced to suffice when the mill added to its capacity. Capacity was again increased, and again the same bin managed to suffice. This was possible because spare fuel piles were kept in reasonable enclosures around the yard. It was cheaper to add fuel to the bin than to build a very large and very expensive new bin, compounding the capital outlay.

DISCUSSION PERIOD

QUESTION: Is there any difference in efficiency between heating with steam and heating with hot water?

BACHRICH: Basically there is none. But because of the legal constraints, there is one very important difference. The temperature of the hot water is limited to 250 degrees Fahrenheit, which limits the temperature in the kiln to about 180 or 185 degrees Fahrenheit. If you could run hot water at a higher temperature, you could get a higher temperature in the kiln, obviously. But you cannot run a high-temperature kiln at 230 degrees Fahrenheit when the temperature of your hot water is only 250 degrees. There is no way to do it. The heat exchanger area would have to be increased significantly, which is too costly, and more fans would be needed to overcome the resistance.

QUESTION: Is there a difference in capital investment between the two systems—steam and hot water.

BACHRICH: The difference is very small. The hot-water kiln, due to its lower temperature medium, will require additional heat coils, but the boiler will be a little cheaper. With steam, a 150-psi boiler will be slightly more expensive, but you will have considerably fewer heat coils in the dry kiln. Do not forget also that with steam, you have the advantage of better humidity control. I realize that with a hot-water kiln you have no waste to the outside, as with a gas kiln, from the products of combustion. Nevertheless, you either inject hot water or you inject cold water. Your humidity level control is not as good–it cannot be as good as with a steam kiln. Nothing is. But if, with regulations as they are, you can save the greater portion of $60,000 in wages per year and you are drying only softwoods (which would permit you to operate on fairly low humidities), you should be all means consider a hot-water kiln. Certainly $60,000 is a considerable sum of money and in a few years it might be $70,000 or $80,000. Wages certainly are not going down.

QUESTION: Do you stop the infeed conveyors with a heat-sensing device in the boiler?

BACHRICH: No, we use a little hopper. I think most feeder installations have a small hopper or surge bin which must be filled to an agreeable level. We have tried in the past on much smaller installations to make the fuel bin a direct and immediate function of the heat demand in the kiln. We have found this leads to many small control problems and requires too many people monkey-wrenching around with it. It is better to feed from an intermediate fuel hopper so there are absolutely no hangups or problems. It is built in a way that insures the fuel will spread rather than contract.

QUESTION: How do you condition lumber in a hot-water kiln?

BACHRICH: You really cannot condition in a hot-water kiln. You can only hope to achieve a reasonable level of humidity. It is much better than a gas-heated kiln, however, because you do not have the constant ingress of products of combustion and the constant exit of kiln air. So a hot-water kiln is, relatively, much better than a gas kiln but still not nearly as good as a steam kiln in terms of conditioning. If the law permits–and it does, for example, in Quebec–you could add a small 15-psi low-pressure steam boiler on the side and use it for low-pressure steam injection only. This is a good method, but it depends on the jurisdiction under which you operate.

There is, incidentally, no sense to these regulations. Hot water at 250 degrees Fahrenheit is a more dangerous fluid than steam at 250 degrees Fahrenheit. If a steam main were to break or if you had a break in the heat exchanger, wouldn't you rather have steam there, which will largely blow into the atmosphere, than solid hot water? Hot water at 250 degrees will not completely flash into steam. Much of it will fall in a solid sheet. The regulations are not based on reality, and it is up to industry to fight them.

QUESTION: Are any hot-water kilns drying at schedules as low as 32, 34, or 36 hours or are they all in the 48- to 72-hour range?

BACHRICH: We have hot-water kilns running on 72-hour schedules drying northern spruce. On the average, according to the figures I presented, a spruce kiln on high pressure steam will work in 48 hours on reasonably high temperatures around or above the boiling point of water. Hot-water or low-pressure steam kilns will work to 185 degrees Fahrenheit and will dry in about one day more. But you have raised an interesting point. If you consider the capital expenditure and the regulations under which you must work and the amount of money you must pay out in salaries to engineers, you have a choice. Instead of drying in 48 hours with steam, you can dry in 72 hours—and this is a comparative figure for white spruce. You are saving the cost of the engineers, but you will have to buy another dry kiln. If you take more time, you need more capacity—or you reduce your production.

With high-pressure steam, you will be able to get the same drying times as with gas. You will not, however, get the same drying times with low-pressure steam or with hot water that you can with a modern gas kiln. It just is not possible, because the heat input is too slow and the ultimate heat you can arrive at is not as great as with gas. With a gas kiln, you can dry to 235 degrees. You cannot do that if your basic fluid is only 250 degrees. You then have to trade off the cost of gas or of butane with the cost of equipment and perhaps with the cost of additional dry kilns.

15

Optimum use of your top resource—people

Jay Gruenfeld, Vice President—Lands & Forestry, Potlatch Corp., Lewiston, Idaho

In the next few years, great advances clearly will be made in sawmill and plywood plant technology. In my opinion, however, this is not the area where the greatest progress can be made toward improved resource use.

Some years ago, shortly after I had been put in charge of a logging camp, I was talking with a fellow named Bill, one of the top loggers on the West Coast. We were discussing the importance of choosing the right kind of equipment for various types of logging jobs when he said, "Jay, you can take your pick of the equipment, but give me the pick of the men and I'll outlog you every time." He was right, of course. Bill not only assumed that *people* were the most important part of the operation, he acted on that assumption.

I mentioned this story to a man with much experience in the plywood business. He told me there was a parallel in the plywood industry. He said he is continually amazed at how often motivated, capable people produce quality products at a lower cost than a more modern, better equipped plant that lacks the right people and real motivation.

I suppose the common goal in most production-type jobs is to achieve objectives through people working with machines. Obviously, I believe people represent the more important element.

Identifying the cause of people problems

I have asked myself: What is the common factor which results in people problems? Consider, if you will, some people relationships and problems:

- Husbands and wives have people problems as a result of jealousies and misunderstandings.
- Students and professors just cannot seem to understand one another and frustrations develop.
- Ministers often talk "over the heads" of their congregations, so are ineffective.
- Trade unions and industry or bosses and workers often have problems resulting from distrust and lack of understanding, even though both may bargain in good faith.
- Public and private agencies often battle because one does not understand the policy restrictions which determine the other's actions.
- International strife developes from misinterpretation or ignorance of actual conditions.

Similar problems affect all of us. Most of the time these problems result from a simple failure to communicate, an inadequate ability to give or receive information. Using words to convey meanings often leads to confusion.

I can almost anticipate a positive response when I ask: Have you ever had serious communication failures between your sawmill people and the loggers? What about between the loggers and the foresters? Or the sawmill folks and your sales people? Or between workers on the green end and the crew on the dry end?

In 1948 I prepared a thesis on a logging and sawmilling operation in the Walden, Colorado, area. The operation was cutting lodgepole pine and small Engelmann spruce and hauling it to a string of portable mills and to a fairly large circular mill at Walden. Later they shut down the big mill. The reason given was that they had not realized how many small logs there would be. The mill simply could not handle them economically. I have heard similar reasons given for the failure of at least three other mills. In each case, failure to communicate the raw material supply—that is, to give and receive information—was the primary cause of the operation's failure.

Improving communications

The company I represent, Potlatch Corp., emphasizes communications, including communication of the major philosophies, objectives, and values of the organization. In fact, the company's philosophy, objectives, and values are published in a booklet that is made available to *all* employees. It is revised on a continuing basis. I will guarantee it includes motherhood statements but they are of a positive nature, I feel. The booklet is extremely useful.

III. COMPANY VALUES

While working to attain the company's stated objectives, a very important set of internally and externally-oriented values will be maintained.

A. Human Resources

1. The company's goals can be best achieved by building an employee group that is highly competent and productive in its work activities, loyal to the company and committed to its philosophy and objectives. To attract and maintain this kind of outstanding work force, the company recognizes its obligations to:
   a. Provide wages, benefits and other conditions of employment that, on the whole, compare favorably with those of competitors.
   b. Strive to provide regular, steady work.
   c. Provide safe and attractive working environments.
   d. Afford employees opportunities for personal growth and advancement to the limits of each individual's capabilities and potential.
   e. Respect the dignity and integrity of each employee as an individual.
   f. Maintain fair and businesslike relationships with all employees, union-represented or non-represented.
   g. Encourage open, two-way communications and understanding on all matters of mutual interest and concern.
   h. Take affirmative actions which fulfill the company's legal and moral obligations to provide equal opportunity for all employees regardless of race, color, religion, national origin, handicap, sex or age.
   i. Encourage individual involvement in civic, community and professional group activities.

• • •

D. Natural Resources

1. Forest and land resources will be managed to provide an optimum economic rate of return to the corporation, consistent with a deep concern for the environment and the company's philosophy.
2. Timberland will be farmed so there is a perpetual crop of raw material.
3. Harvesting and other land management techniques, as well as all planning and operations, will be conducted with a high degree of awareness of present and future impacts on the public and the necessity for good stewardship.
4. Forest lands will be open for public recreation except when access must be denied to protect employees, property or the public.
5. Energy conservation of all kinds will be vigorously pursued.

Excerpts from Potlatch Corp.'s booklet (distributed to all employees) describing the company's business philosophy, objectives, and values.

New employees have a much better chance of knowing what the company stands for when the information is readily available in a booklet. Many organizations could benefit from such a widely distributed, easily read booklet. Our booklet, so far as I know, is unique in the forest products industry.

When I was interviewed for the job I now have, I was introduced to the booklet immediately. I was quite impressed with it. I knew the president of the corporation had come from Mobil Oil Co. And I thought to myself, "Those oil people make so much money that perhaps this sort of thing is common in that industry." My seventh interview was with the president and he asked me quite a few questions. The first chance I had to ask him anything, I said, "Mr. Madden, is this sort of booklet common in the oil industry?" because I had been told Madden had had quite a bit to do with the booklet. He said, "No, I never saw one before in my life." I was impressed that he believes communications are important enough, for instance, to take the potshots that are bound to come when you put out something like this. It can be quoted out of context; it can be used against you. But to Dick Madden, it is worthwhile to have this material published. Most people at Potlatch feel the same way.

I would like to quote a few things from the booklet.

- Be sensitive to the needs and desires of employees.
- Encourage open, two-way communications and understanding on all matters of mutual interest and concern.
- Encourage individual involvement in civic, community, and professional group activities.

It takes some courage to put out something like this booklet, but it is a plus. When a new employee comes to the corporation, he can learn so much by just spending 15 minutes reading the booklet.

Now for a somewhat critical comment about the Potlatch Corp. Appropriately enough, it came from our Communications Council, a critical board of review consisting primarily of people from outside Potlatch. We asked members of the Communications Council to give us their opinions on how we might improve what we felt were relatively good community relations. They felt the key people we should communicate with were our employees, on a two-way basis. The people of Potlatch could then go out into the community and communicate the facts to other people. The council saw this as being more important than the normal advertising and other public information programs. These people—the employees—have credibility. They are the voters; they are the shapers of the things important to all of us.

Industry must educate nation's voters

So far I have talked about people on the job. Let us take the concept a step farther and look at an even bigger picture. Under our system of government, people do, in the long run, get pretty much what they ask for. For example:

- It is people who set the standards that determine optimums.
- It is people who ultimately determine private property rights—and that involves land-use planning and zoning, for example.
- It is people who produce the goods and services that satisfy their wants and needs.
- It is people who shape the political climate which determines what the tradeoffs will be, for instance, between jobs and recreation.

The last point is especially important today, since the forest products industry faces a national crisis that could close many mills, resulting in the loss of thousands of jobs. In fact, from a historical perspective many believe we are today facing the most critical challenge the forest products industry has seen in more than 30 years. This crisis was created by a legal decision involving the Monongahela National Forest in West Virginia. The Chief of the U.S. Forest Service has said that, applied to all 155 national forests, this decision could reduce the national forest timber harvest by 75 percent in 1976 and by 50 percent for the rest of the century.

It now appears that Congress must pass new legislation to avoid such a catastrophe. Congressmen will listen to the people who elect them. You will be receiving information from your companies on the Monongahela problem. It is very important that you help communicate it. People act according to what they believe so it is important that we all get the facts.

People act according to what they believe—and they do not necessarily base their beliefs on the facts, particularly if those facts are not made available to them. You might have a fellow on the job who is a pure cull—but if people do not think he is a cull, he can create plenty of havoc because people act according to what they believe about him. I feel confident that if the people of this country knew what the tradeoffs were in the land management issues—if they really knew the facts—the industry would get a much better shake in the choice between jobs and, for instance, preservation of areas for a single use. Multiple use would be a much higher priority if people knew the facts. But only some of the people know them. It is up to industry to do a better job of communicating with the voters outside the industry.

People tend to support what they help create. We talk about participative management and things of that nature. If you should decide, for instance, that some function in your mill calls for an operating booklet or manual, there is a much greater chance that it will be followed and that it will be effective if it is created with the participation of the people who will act under its guidelines.

In most situations, improved communications would increase productivity if the communications program were developed with full recognition that

- People act according to what they believe.
- People tend to support what they help create.
- People are the top resource.

DISCUSSION PERIOD

QUESTION: One of the issues facing the industry, at least those companies with large government contracts, is putting women and other minorities into executive positions. The industry as such has not made too much progress in this direction. As an executive recruiter for the forest products industry, it is a problem I deal with daily. Why do you think the forest products industry is behind other industries in this respect? What has been your personal experience in promoting women to executive jobs? Do you think it is beneficial or detrimental in terms of productivity when women fill these positions?

GRUENFELD: One reason, certainly, is the problems associated with making any kind of a change. These are not unique to our industry. Whenever you change a long-standing practice, you run headlong into resistance to change. I think we have made some real strides. Most corporations are making honest attempts. But resistance to change and the fact that our culture does not think of putting women in manual labor mill positions are major factors. It is difficult, therefore, for women to get the background which creates the opportunity for advancement. Another point, in terms of woods jobs, is the physical factor. Women can do much more physically than has been realized, but it is a change from a historic cultural pattern.

16

Human factors in engineering design

Joe G. Sublett, Manager-Industrial Design & Human Factors, Weyerhaeuser Co., Tacoma, Washington

This discussion will be about people—what we know about them and what we do with them.

Since the beginning of organized work activity, man has been expected to fit into his occupational environment. The 1876 Chicago World's Fair motto, which follows, aptly characterized this philosophy: "Science finds, industry applies, man conforms." This unfortunate attitude has, for the past 100 years, caused great problems for industry and wrought considerable physical and psychological stress upon the worker. Since most work stress results from continued exposure to poorly designed, illogical, and hazardous work spaces and procedures, many workers gradually became permanently disabled or suffered permanent harm without the cause being obvious. This resulted in problems. Worker motivation degenerated. Absenteeism, job turnover, and union dissension grew. And efficiency, quality, and productivity suffered.

Exactly why this situation developed is not clear, but during this period of rapidly developing technology, great emphasis was placed on scientific and engineering disciplines. Accurate data about people (how they see, hear, react, think; how big or small they are; how far they can reach or bend; how strong they are) have been developed only in the past 25 years. The engineer could not be an expert on all these "human factors," and industry recognized the need for new interdisciplinary fields such as industrial design and human engineering.

Automation increases importance of man

With the advent of modern technology and automation, the mistaken idea that we need not place much emphasis on the operator has arisen. Nothing could be further from the truth. The more sophisticated and automated our equipment becomes, the more critical man's role and the more costly man's mistakes. Man is the central reason for design. Automation is simply an extension of man's capability—and is therefore designed for man. This concept is referred to as "designing from the man out." That is, we start with man and provide the accessories he needs to carry out or reach a prescribed objective. This results in the most practical, common-sense, simple design to meet the system objective. In design, the simplest solution is always the best.

The most effective way to apply human factors in industry is to utilize the Systems approach—or systems engineering, as it is sometimes called. By "system" we mean simply a group of components or subsystems designed to work together for a common purpose—that is, an automobile, a space vehicle, a sawmill. We generally refer to this system as a man-machine system. Man is thus a system component. Consideration of the human operator's capabilities and limitations provides insight into the ways in which he can best be utilized as a component in a man-machine system.

Man's function in the system

As a receiver. Proper inputs to the senses (seeing, hearing, touch, and so forth) result in man's being a highly reliable receiver. His rate of perceived input seems to peak at about 10 successive items per second. His input capacity is easily saturated. Care should be taken to avoid inputs that are overlapping, competing, or visually and orally incompatible.

As a computer or evaluator. Man is an exceptionally good evaluative computer. Man is able to make decisions based on past experience. He also has anticipatory capabilities. He is the only computer able to solve problems by logical induction. For example, hovering a helicopter is a task man can perform quite easily. Done by a computer, however, it requires the simultaneous solution of nine equations.

As a transmitter or control mechanism. Man's outputs are motor responses—such as operating control buttons, levers, foot pedals, and so forth—are relatively slow and low-powered. For steady-state conditions as a controller, his output in highly developed skills (typing) seems to be a maximum of about 25 bits per second. The rate for unskilled tasks is about 2 bits per second.

Man is very flexible and can perform well in many different jobs if his limitations are not overlooked. Use machines to relieve the man of as many routine jobs as possible, but use the man to supply the judgments and flexibility which machines cannot.

Man's performance usually tends to deteriorate as a function of time on the job. This is seldom a result of physical tiring per se. It results, rather, from boredom, inattention, and lack of motivation. Care should be taken to design a man's job so he is not forced to operate near his maximum load limits for very long. In addition, control over his ambient environment must be exercised. The cumulative effect of several environments interacting at the same time (noise, temperature, dust, and vibration, for example) will significantly reduce man's tolerance to any single environment.

Let us look at a typical man-machine system (Figure 16.1) with which we are all familiar. We have the machine—the great mechanical-electrical-hydraulic monster—and we have that mysterious thing referred to as the black box. Then we have the operator. He is often placed in a completely illogical control booth (usually a plywood shack stuck up as an afterthought), with poor vision of the operation. He is supplied with a completely illogical control panel with wide rows and columns of multicolored, multilighted controls and displays. I call this the Christmas tree effect. There is generally great misuse of color on the control panel as well as lack of well-designed geometry for the operator's seating and vision. We bombard him with tremendous noise, dust, and vibration. Unfortunately, this is not far from reality in many mills. There is a great need to consider man, not as an afterthought, but during the design of the system.

This word of advice to mill owners and operators: Do not depend on your equipment suppliers to cover the human factors. And a word of advice to equipment suppliers: Do not assume that

Figure 16.1. Ignoring operator's needs and limitations in engineering design has resulted in "man-machine" systems such as this.

mill operators will continue to accept illogical, hazardous machines and operator work stations. There is a great need for sawmill industry people and the suppliers of sawmill machinery and equipment to work together to create a people-oriented man-machine system.

The modern trend toward automation and pushbutton tasks has not decreased the need for proper man-machine design. In fact, it serves to emphasize the importance of effective design since human control of such automatic systems has a more direct effect upon a much larger array of equipment.

To cope with the diverse problems which are encountered in man-machine design, it is necessary to draw upon several widely different technical specialties, including:

Medicine	Psychology	Anthropology
Biology	Physiology	Operations research
Physics	Mathematics	Acoustics/Optics
Graphics	Electronics	Industrial hygiene

Computer technology   Mechanical engineering

Through the years of their development, these fields have become increasingly specialized in language and techniques, and thus have placed a difficult burden upon the designer who would attempt to unify the results. Recently, however, the rapidly emerging professions of industrial design and human factors engineering have begun to integrate the knowledge of engineering technology and life sciences.

Defining the terms

Industrial design is the design and engineering of environments and equipment to be compatible with human requirements, capabilities, limitations, health, safety, and comfort. Industrial design includes the development of systematic methods for considering man in the design of systems and the application of human factors and methods throughout design.

Human factors is the scientific determination of physiological, sociological, and psychological facts which affect human behavior. Human factors include the development and application of procedures and principles for the design of work environments, work station geometry, consoles, seating, instrument panels, controls, and displays. Human factors considerations are also included in maintainability design, environmental control, safety, work procedures and task analysis, training, and measurement of human performance.

One obvious physiological factor is man's senses. Noise, dust, vibration—all have their effect on man's senses. Another physiological factor is body dimensions: the height of the operator's seat and the degree of adjustability that should be provided at an operator's station. We must do away with the myth of designing for the so-called average man. There is no average man. If we designed for the average, our design would not fit the 49.5 percent of users above and below the point to which we designed. Good human factors design includes providing a range of adjustment. A number of mills are using women operators in various areas, so they realize the need for adjustable seats and control stations.

Sociological factors include cultural influences. We read from left to right, top to bottom. Certain colors and symbols have meaning to us. We expect certain movement relationships. If we turn a control knob clockwise, for example, we expect an increase;

counterclockwise, a decrease. Understanding these basic facts helps us do a better job of designing operator work stations. In so doing, we also eliminate many operator mistakes.

Psychological factors is an area of misunderstanding—and it is a difficult area. It includes motivation. An old mill joke is the use of a 2x4 to beat and threaten the worker into submission. We know that people will do much, much more and will do a better job of it if they are properly motivated. The worker must be provided with job enrichment, job value. People want to contribute. They want to get some satisfaction from their work. Management people feel that way, so why shouldn't the worker on the green chain or the equipment operator feel the same. Stress and boredom are problems with which each of us has to contend, so we must accept that the people who work for us have the same needs and problems. We must learn how to deal with those problems.

I am continually reminded that the sawmill industry is not like the aircraft or aerospace industry—that human factors are not needed in this industry, that a sawmill is not like a 747 airplane. My answer is that the common denominator in these industries is people. The man has not changed simply because you have taken him from an airplane and placed him in a sawmill. His basic needs and limitations are still the same.

The "arm-chair" console (Figure 16.2) is a simple, but effective operator station design. The console is designed for good visibility and comfortable seating geometry. The seat has an adjustment range of 4 inches both vertically and horizontally to accommodate operators of various sizes. The panels are designed to provide good visual contrast and controls are arranged in a functional sequence. Panels hinge open for easy access to wiring.

Basis for human factors design

Industrial design/human factors design decisions are based, to the degree possible, upon explicit information regarding user requirements. Operational design features are considered in terms of the desired performance under the intended operational environment. Thus, the criteria for the application of industrial design/human factors design considerations are the optimum use of both man and tasks to maximize man-machine effectiveness.

Figure 16.2. Design of this operator console takes into account the human factors—physiological (operator size), sociological (expected sequences and relationships), and psychological (comfort).

The primary objectives of the industrial design/human factors efforts are to identify and explore technical problems relating to the man-environment or man-machine interface and to provide design solutions to these problems. Industrial design/human factors design efforts will follow a parallel, but interactive development cycle with engineering design. These design efforts should thus be an integrating force and provide a framework for the balancing of man-environment, man-machine requirements at system, subsystem, and component levels.

What is the scope of human factors engineering?

The overall scope of industrial design/human factors engineering includes five major areas of contribution:

1. Systems/operational analysis
   - Process objective
   - Man/machine functions
   - Input, process, output requirements
   - Control/display requirements
2. Operator environment design
   - Noise/vibration
   - Temperature/humidity/ventilation
   - Lighting
3. Work station design
   - Vision and reach geometry (standing/seated)
   - Control consoles and panel layout (sequence)
   - Control/display types
4. Operational safety
   - Equipment layout/hazards
   - OSHA requirements
   - Safety procedures
5. Operator training
   - Operator requirements/skills
   - Training plans
   - Motivation/performance

The ultimate result of applying human factors to engineering design is increased operator motivation, efficiency, and productivity. This study of the role of human factors in engineering design leads to these conclusions:

1. The limitations of human factors applications are associated with their scope, cost, and acceptance. Cost should be examined in light of potential gains.
2. To avoid the mistake of considering people as isolated beings, human performance should be investigated within the context of the system in which it exists. The systems concept provides a basis for considering all critical factors that may lead to improvement.
3. Although human factors engineering is a relatively new field, a great deal of information has been developed pertinent to human performance. This information may be adopted for direct application to man-machine systems design.

17

The impact of land-use planning

V.M. "Whitey" Howard, Vice President & General Manager, Seneca Sawmill Co., Eugene, Oregon
William Lesh, Assistant Director of Public Relations, Publishers Paper Co., Oregon City, Oregon

Part one by V. M. "Whitey" Howard

Land-use planning on federal Forest Service lands might more appropriately be described as the arbitration between the takee and the takor, or who gets the benefits and who pays the costs.

Public lands belong to all the people and must serve all needs. No longer do huge undeveloped lands remain. The name of the game today is *management* of our existing resources for best production now and in the future.

What interests do we have in public land-use planning? Many of us in the West depend upon wood products from the public forest lands for our livelihood. We must also depend to some extent upon public forest lands for our recreational needs, since more than half our lands are owned by the federal government.

Forest Service land-use plans being developed in the West show a pattern of reduced fiber production—approximately 20 percent below current harvest levels. Such a reduced production means fewer jobs, higher taxes, and greater wood imports to provide the housing needed for all our people.

The total Forest Service Region VI actual harvest has averaged about 5 billion board feet per year. According to Dr. Adrian Gamache (a Portland, Oregon, economist) the 20 percent reduction suggested by the Forest Service's alternative would mean 22,110 fewer jobs, a $270 million reduction in payrolls, and a $50 million loss in in-lieu payments to the counties. It would also reduce the available logs by 1 billion board feet. These figures are based on a $200 stumpage cost and current wage rates.

Those of us in private industry, the diminishing part of our economy, must set and meet realistic production and profit goals to stay in business. This is not true of the U.S. Forest Service. Neither the Administration nor Congress has chosen to set forth quality and quantity goals for the national forest system. Planning, then, is a political football game between the biological ability of the forest to produce and the projected needs of the people.

You will see in Figure 17.1 that the demands of the people, plotted by the top line, will greatly increase. The middle line represents current forest production. The bottom line represents what appears to be the direction taken by the Forest Service in determining allowable cuts under new land-use plans.

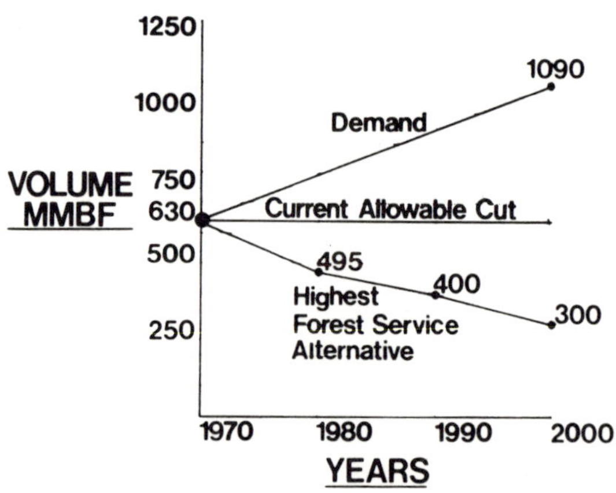

Figure 17.1. Demand, current forest production, and direction of U.S. Forest Service land-use planning to the year 2000.

Note definitely that once a Forest Service land-use plan is finally adopted and clear of the courts, it can be changed only by a completely new environmental impact statement and procedure. Thus, the plan virtually becomes cast in concrete. We will be forced to live with it for some time—a minimum of 10 years.

Land-use planning and the Willamette National Forest

I would now like to examine the Willamette National Forest land-use planning process to date.

Oregon's Willamette National Forest is the largest producer in the national forest system. It has about 1.7 million acres and provides logging, range, watershed, wildlife, fish, and all outdoor recreational pursuits, as do most national forests in the West.

To prepare a plan, inventory was taken of everything from trees, soils, and water to campgrounds, wilderness, recreation, and so forth. This information was fed into that magic black box called a computer. The output was then arranged into several series of combinations, or alternatives.

One alternative started with 1.7 million acres. Of that, 15 percent was designated wilderness; 5 percent water, roads, and nonproductive areas; 3 percent designated for recreation and winter sports; 12 percent undeveloped, roadless, and administrative site areas; two levels of scenic constraint totalling 30 percent; and the remaining 35 percent was designated for constrained management due to visual impact. Note that not one acre was designated for pure intensive timber growth without constraints.

The first maps prepared by the Forest Service indicated approximately 40,000 acres in a nonallowable cut category, because they were high-elevation lands which could not be reforested within an acceptable time. Examination of the Forest Service's own reforestation records, however, revealed 95 percent successful reforestation on clearcut units inside that category of land classification. Later, two other examples of questionable classification were discovered.

One example was in a land-use designation called "special interest," which withdrew up to 138,000 acres from fiber production. "Special interest," we understand from Forest Service definition, is a classification to protect unusual recreation and scenic values in a substantially natural condition, with timber harvests and roads not permitted. The Loggers Butte area is a misclassified area. It had already been logged, reforested, and has roads. It did not, by industry analysis, meet the Forest Service requirement for the "special interest" classification.

Another example is a timber management classification of approximately 140,000 acres as "marginal" due to soil conditions, slope, and other natural forces. No consideration was given to helicopter or balloon logging, which have been used successfully in these areas for nearly five years.

Note that I have been talking about basic areas of land allocation. Let me make crystal-clear that there are two fundamental parts to the allowable cut calculations, each entirely separate from the other. The first is the acres allocated; the second is the intensity by which those acres will be managed. If acreage is not first allocated to fiber growth, *no* intensity will be practiced. If acreage is allocated to fiber growth, the amount of yield will vary with the degree of management intensity applied.

When comparing various land-use planning alternatives, be absolutely sure that you use *current* funding or *current* management intensity as the guide to intensity level for each alternative examined or proposed. Only thus can you obtain an apples-to-apples comparison. The allowable cut output by itself is a grossly inadequate basis for comparison. After a land-use plan allocates acres as not designated for fiber production, they are lost.

We have all heard so-called conservationists or environmentalists calling for acreage withdrawal for wilderness on one drainage and increased intensity of management on another drainage as a bribe to gain support for their position. I have not heard a guarantee of funding, however, from these people.

My emphasis is directed toward fiber production. The Willamette National Forest made the same detailed analysis and projects for *each* of the uses. Some are compatible; some are not.

The Forest Service made a table showing all the uses and their rating of the uses or multiple-use compatibility. Sound simple? Not so. The study required 680 pages in published form.

To give an idea of its complexity: The Forest Service delineated about 200 land tracts and 12 land-use designations. Throw those into a Mixmaster and you can visualize millions of possible combinations. Further complexity arises if three or four levels of management intensity are tossed in for good measure.

After much shifting of land classifications and numerous calculations of cost/benefit ratios, allowable cuts, recreation uses, and so forth, the Forest Service offered seven alternatives to the public. These are included in the *Willamette National Forest Draft Environmental Statement.* Two of the alternatives were labelled by the Forest Service as being impossible or not viable under current laws and regulations. With very minor changes (a matter of 2,000 acres) the two nonviable alternatives would have increased allowable harvests and been within the current laws.

The five viable alternatives were submitted to the public for review and were outlined at a series of meetings around the Willamette area. The public was invited to respond within 90 days.

With a slide show, brochures, cards, and petitions based on fact and professional studies, the Western Environmental Trade Assn. (WETA) and the Willamette Timbermen appealed to the public to understand the impact of the proposals upon jobs, taxes, and housing. They also presented an alternative proposal that would have increased the forest yield by approximately 4 percent, that would have doubled developed campground capacity, that would have provided for a new ski area, and that would have provided for 26,000 additional acres of dispersed recreation.

The environmentalists also went to the public with an alternative. That alternative would have reduced fiber production on the Willamette National Forest by more than half.

Analysis by the Forest Service of public response to the *Draft Environmental Statement* was done by a process called Codinvolve. This indicated 2,430 responses. Happily, only 182 responses favored the Sierra Club coalition alternative and 32, the Forest Service alternatives.

The response overwhelmingly favored jobs, reduced taxes, a vigorous economy, and the Timbermen-WETA alternative. The Forest Service analysis did not openly acknowledge receipt of a referendum-type petition with about 4,000 names that favored the WETA alternative.

The Forest Service Codinvolve analysis made little, if any, recognition of two professional documents submitted on behalf of WETA and the Willamette Timbermen. The first of these was

Planning and People, an 85-page economic analysis of the *Willamette Draft Environmental Statement* by Portland consulting economist Dr. Adrian Gamache. *The Willamette National Forest Draft Environmental Statement: Land-Use and Timber Management Alternatives–1975*, prepared by Dr. Carl Newport, of Mason, Bruce & Girard, consulting foresters, Portland, Oregon. These two studies contain excellent analyses of economic yield and allowable cut, and clearly point out errors in the *Willamette Draft Environmental Statement.* Furthermore, they make positive recommendations to the Forest Service on actions to be taken prior to the issuance of a final environmental impact statement.

There is no way to judge at this time what effect the public input or the WETA-Timbermen input will have on the Forest Service's final environmental impact statement. But be assured we will lose by default if we do not get to the public at large.

As James Russell Lowell said, "The pressure of public opinion is like the atmosphere. You can't see it, but all the same, it is 16 pounds to the square inch."

We must get to our own people and to the general public to make public opinion visible. We must clearly explain who gets the benefits and who pays the costs.

Part two by William Lesh

I intend to concentrate on land-use planning in Oregon, but in Washington and California similar legislation is in the works. Two bills now active in California may have been exported from Oregon. They are extremely similar to Oregon statewide land-use bills. California also has a coastal zone management proposal that will make existing coastal zone commissions permanent. Another will plan for agricultural lands. So Californians should be alert.

Washington has its own coastal zone management act. That is something to watch because the Washington legislature is already dealing with a statewide land-use plan.

There is also a major movement for national land-use legislation. Find out in this election year where the congressmen and senators up for reelection in your area stand on land-use planning. We do not need another federal agency telling the states how to do a job. They can do much better on their own.

Understanding plannerese

To set the tone for this discussion, I offer a quotation from one of Oregon's regional planning agencies. This agency covers some one million people around Portland. The list to follow details some of the agency's regional goals. You may not understand them—neither do I. They sound a little like the Declaration of Independence, but remember, they represent a planner speaking.

1. Fulfill the basic food, housing, clothing, and health needs of all the region's citizens.
2. Assure the personal safety, continuing income and employment, and safeguard the properties of all the region's citizens.
3. Promote freedom of choice in personal endeavor and way of living consistent with the rights of others.
4. Preserve or conserve nonrenewable natural resources.
5. Preserve or conserve renewable natural resources.
6. Preserve, conserve, and create human-made resources.
7. Integrate and apply goals through standards and procedures for planning.

Any plausible explanation of that last goal would be welcome!

Probably the basic problem in land-use planning is the lack of understanding or the split between the laws and the property owners or average citizens. Planners talk a language different from ours. I call it "plannerese." They also use "legalese." The courts become involved. You hear acronyms such as LCDC, OCC&DC, CZM, CIAC, CAC, CRAG, ODOT. You hear terms such as *comprehensive plan, framework plan, the Baker decision, the Faisano decision.* Those are just examples of words and acronyms you will encounter. I will try to avoid those terms, but the complexity of land-use planning does not make it easy.

Do not feel you can assign the land-use planning issue to one man in your company and say, "That is Lesh's area. It says here under his name 'land-use planning expert' so I don't have to worry about it." That is the first mistake—and it is a mistake many of us in the wood products industry and in many other industries have made in the past. We have made the same mistake in our private lives. That mistake is being used against us by a number of organizations—I will not classify them all necessarily as environmentalists, but they seem to have that tendency.

So your first assignment—and it is only the first—is to find out what is going on in your area.

Organization of land-use planning in Oregon

I mentioned the acronym LCDC. In 1973 Senate Bill 100 was passed in Oregon. It set up a statewide land-use planning organization called the Land Conservation & Development Commission (LCDC). At the same time, a lesser known act, Senate Bill 769, also passed. It set up something called the Columbia Region Assn. of Governments, known as CRAG. The regional goals quoted earlier are included in CRAG's draft framework plan. The legislation setting up the LCDC told the commission its first job was to set up a series of goals—policy statements, if you will—that would be applicable to every city, county, and region in the state. Every city and county must meet LCDC's goals and guidelines.

This brings us to the second assignment. If you do not have a copy of LCDC's newspaper that lists the 14 adopted goals and guidelines, contact their headquarters in Salem for a copy. Each city and county in Oregon is required to deal with those 14 goals and guidelines in its comprehensive plan. By the end of 1976, each city and each county in Oregon (276 local governments) must have a completed comprehensive plan or must have set up a timetable for completion of a plan.

What the LCDC guidelines include

Let me give you an idea how far LCDC's land-use planning guidelines extend. They cover things like citizen involvement. They tell local governments how their plans must be set up. For example, when dealing with agricultural lands, EFUs (exclusive farm-use zones) must be formed. These require the preservation of agricultural land, and I use the word *preservation* in its strictest sense. Preservation means that land in an EFU zone *cannot* be used for any purpose other than farming. It does not matter whether it is economical to farm that land; it must be zoned "exclusively farm use."

Turning to forest lands—we are not *preserving* forest lands, we are only *conserving* forest lands. Open spaces, scenic and historic areas, and natural resources are treated in another goal. So are air, water, and land resource quality. Some believe the LCDC should be the overall state agency that can tell local governments and other state agencies how, when, and how much they must spend on air and water pollution control equipment and on improving air and water quality.

You must find out the location of *flood plains*, another planning term. Recreational needs, the economy, and housing are treated in the goals. Public facilities and services, transportation, energy conservation, and urbanization are also covered. As part of the urbanization guideline, an *urban growth boundary* must be drawn around every incorporated town in Oregon to limit the growth of the community. That urban growth boundary also limits where whater and sewer lines go, which limits where you can build a house, how many houses can be built, or where you can build a sawmill or a plywood plant. It involves many other limits as well.

Comprehensive plan is law

I will not even attempt to define the term *comprehensive plan.* I will, however, tell you what the term meant in the past. It was once a general document that set goals for an area for the future. It was an extremely flexible document. It was not a document you consulted if you wanted to make a zone change or to determine whether you could or could not do something in your area. Now, thanks to a court case referred to as the Baker decision (*Baker* v. *Milwaukie*), the comprehensive plan is the law. Zoning must follow the comprehensive plan. Where the comprehensive plan specifically states a guideline—for example, there shall be no more than X units per acre in a particular area—zoning must implement it.

The Milwaukie case involved a zone change. The city of Milwaukie allowed more apartment units to be built on a piece of land than the comprehensive plan allowed. Without a recent out-of-court financial settlement to help the young woman who started that case, those apartments would have had to be torn down.

That is a serious situation. There are many similar examples of violations of comprehensive plans. A number of court cases in Oregon now state that those things which have been built or done in violation of a comprehensive plan may have to be torn down. The plan is king. The plan is the constitution. I am stating this as strongly as possible to urge you to find out what is going on in planning in your area.

Identifying boundaries

What are some other dangers involved in land-use planning? One is this: You will find a map in every comprehensive plan. Planners have a strange disease. I hope it is not communicable, but it seems to be universal among planners. I call it "blobitis." Comprehensive plan maps usually are not much bigger than 8½ inches by 11 inches. (Occasionally they are printed on legal-size paper.) They are generally covered with little dots in pretty colors. You will also find a number of small circles and quite a few wavy lines. These wavy lines are put through the countryside without regard for the legal boundaries they cross. Then the planners list what can and cannot be done within each of the colored areas.

You, then, must know whether you are "blobbed in or blobbed out" of a particular area. That may sound a little strange, but I suggest you ask your county planner what the "blob" covering your sawmill looks like. I am familiar with one situation in which one side of the mill site falls within a "heavy industrial" blob. The other side of the site, where a part of the sawmill is located, is in an "agricultural land" blob. You had better be sure that the applicable blob covers your entire mill site and any other land on which you want to build, expand, or store logs—or else you may not be able to do any of those things. *Remember:* The zoning must follow the plan, not the other way around.

Economic considerations ignored

Another danger in land-use planning is, to use the planners' phrase, insufficient economic input. Perhaps the greatest weakness in most comprehensive plans—and in planning in general—is the failure to plan for economic growth. We plan for natural areas and we plan for open space, because everyone wants open space. We keep industry out of everyone's neighborhood. In fact, one problem in Portland is that the neighborhoods are so well organized they cannot come to an agreement on where anything should be built that affects more than one neighborhood. Thus it seems that Portland will have no new roads, no new industry, and no new high-rise housing because no neighborhood wants these things within its boundaries.

There is little or no economic input into land-use planning. Again, that is where you come in. You must talk to the planners and to the planning commissions and tell them they must plan for more jobs and for more housing. More people are moving into Oregon than are leaving. We have a high unemployment rate. If planning does not deal with this, the state will not be able to afford its enjoyable open spaces.

What else should you watch for? You may think comprehensive plans are a big enough problem to keep you occupied for some time, but there is yet another area for concern.

Greenway under LCDC jurisdiction

The Land Conservation & Development Commission also has set up something called the Willamette River Greenway. This is a special interest area. The state legislature passed a law setting aside both banks of the Willamette River for a "greenway." If you are in an urban area (such as Publishers Paper Co.'s mill at Oregon City) you must consult the city each time you want to do anything other than very routine maintenance within that area. If, for example, you want to extend a building 3 feet (which we are doing), you must get a conditional use permit. It is a long, involved process and must include a public hearing. The permit must be approved by the Department of Transportation in Salem. And if anyone does not like what you are doing, it can be appealed to the LCDC.

You can imagine the problems in rural areas. Those of you with sawmills in rural areas along the Willamette River Greenway that are also in the flood plain of the Willamette River should check immediately with your county and find out what is going on. The Greenway can be extended beyond 150 feet from the low water line to include the entire flood plain surrounding and including your mill. This is something you must become involved with, and which you must watch out for. Your local planning commission is working on this. Find out what it is doing.

What is happening on the Coast?

Let us turn to one other area. A group called the OCC&DC–the Oregon Coastal Conservation & Development Commission–spent approximately four years and a sizable amount of state tax money plus quite a bit of federal money to develop a series of coastal goals for the coastal zone management plan for Oregon. Western Environmental Trade Assn. (WETA) did a study on that plan, which incidentally has been referred to as a horror story. After WETA presented its study to the LCDC, the LCDC quickly rejected all the OCC&DC work. The OCC&DC plan would have eliminated at a minimum–and this was in the good areas–50 percent of all timber harvesting on land from the crest of the Coast Range to the Pacific Ocean. Half that land would have been locked up in areas not designated for timber harvesting. WETA found that in some areas, such as in western Lane County, the figure was as high as 80 percent.

A part of the coastal zone goals has to do with a shoreline zone. The shoreline zone covers up to 500 feet on either side of an estuary and 200 feet on either side of lakes, rivers, or streams. (No definition is given of the size of a "stream.") In addition, any slope of 25 percent or greater is included in the shoreline zone. That includes, in western Lane County, 60 percent of the forest land and would probably preclude clearcutting because of visual and esthetic values. The situation is the same, if not worse, in California and in Washington.

A new look at zoning

Most of us believe we know what zoning is. It tells you what the setback should be on a house you may build or when to change from "agricultural" or "general use" to "heavy industrial" zones.

In Salem, the Oregon capital, there are 2,400 conflicts between present zoning and the city's comprehensive plan–and the comprehensive plan is law. One young man bought a store in Salem and was doing fairly well. The store had a 50-foot setback, which was legal. He went to the proper authorities to get permission to put an awning on his store. He was told he could not put up an awning because the setback had been changed from 50 feet to 60 feet. They were probably going to widen his street and they hoped he could find another store in which to sell his goods because they were going to take 10 feet off the front of his building. All this, without having to give him notice.

That is one example of the problems that exist in zoning. It is not required that you receive notification. Items called "design review ordinances" are a part of zoning. These can tell you that you must plant trees or must set back your building 100 feet, for example, from the edge of your property. These ordinances can tell you what kind of trees you must plan, how fast they must grow, what color you can paint your building, whether you can have windows, what shape those windows can be—and on and on. Again, it is a problem of law and understanding zoning and related ordinances.

The Western Environmental Trade Assn. (WETA) funded an intern to do a study of the Land Conservation & Development Commission's goals in Washington County, touted as the greatest example of land-use planning in Oregon. The final paragraph of that study really sums up what we all have to deal with:

> The LCDC experience is a classic example of citizens and politicians alike expecting more from governmental policies than is possible. Land uses are determined by economics and social forces as well as by governmental policies. In fact, governmental regulations frequently reflect those forces rather than control them. With this in mind, it is only logical to expect major incompatibilities to emerge between state land-use agencies, local governmental units, and private citizens. The LCDC experience has to date been more effective as a negative control on undesirable development than as a positive stimulus to the creation of good environment.

What can you do—as an individual and as a member of the forest products industry? Your mission, should you decide to accept it, is this: Find out what is going on in your area, at the city and county levels. Get your name on the mailing list. Participate in what the planners are doing. Involve yourself in the planning process and provide your own input and the input of your friends and neighbors.

Remember that planning should be as flexible, as simple, and as understandable as possible. You are the only one who can make it that way. A planner will not do it for you.

DISCUSSION PERIOD

QUESTION: What is the typical background of most professional planners?

LESH: I cannot honestly tell you. Landscape architects become planners. Architecture school graduates become planners. The University of Oregon has a department of Urban Studies, which is designed to produce planners. Very few come from industry. We need the input of someone who has run a sawmill or who is a member of a union and works on a green chain. Planners seem to come from all walks of life. The only common bond I can identify is that they are rarely in touch with the reality of the community.

QUESTION: What is necessary for a comprehensive plan change?

LESH: To meet the Oregon LCDC requirements, it is a rather extended process. You begin by making your proposal to the planning commission in your city or county. You must go through a series of public hearings to meet the LCDC requirements for citizen involvement and the planning goal. These public hearings can be as few or as many as the locality desires, so long as no one goes to the LCDC and says they were not handled correctly, which is a distinct possibility. Then the planning commission will have to approve the proposal. Next, the governing body will have to vote on it. At any point, but especially after it has been voted on by one of the governmental agencies, it can be appealed to the LCDC. There are many appeals now active. It is not a quick process. That is why I urge all of you to become involved now while you still have a chance to get your oar in before these things become set in concrete. And that is exactly what they become, even though another LCDC requirement specifies a review phase. There is very little likelihood you will get any major changes in a comprehensive plan without working, I would say, for at least a year.

HOWARD: In Oregon a man by the name of L. B. Day is chairman of LCDC. He signs the first comprehensive plan. Your appeal also goes back to him, and not beyond. He is judge and jury. There is no recourse beyond him.

QUESTION: I understand that we will need approximately one-third more jobs, houses, and services to supply the children we now have in Oregon who are not yet old enough to be primary consumers in those markets. If that is the case, do the plans being prepared allow for that growth capacity?

LESH: In most cases growth—*especially* economic growth—is not a subject that is adequately dealt with.

QUESTION: I am presently serving on the industrial advisory committee in Lake Oswego. One of my tasks is ensuring that our recommendations conform to LCDC and to CRAG. What pitfalls will I encounter?

LESH: The tricky thing about the Portland urban area is that all city and county plans must meet the CRAG goals and guidelines. Then the CRAG goals and guidelines must meet the LCDC goals and guidelines. The LCDC goals and guidelines are fact. CRAG, however, has not finalized its goals and guidelines and does not expect to do so for at least a year or more. This puts you and your locality in the rather tricky position of developing your own plan within the guidelines of the LCDC while having CRAG watching over your shoulder. It may decide a year from now that it does not like what you have done because it does not agree with what CRAG has done. The CRAG situation is a strong and rather frightening example of the problems in setting up a regional body to oversee local governments and to act as a feedthrough to state government. I cannot give you any easy answers. All I can suggest is that you try to get a CRAG planner to hold your hand all the way—which could in itself be quite an experience—and hope that LCDC approves what CRAG does.

18

The national forests: some neglected costs

Barney Dowdle, Professor of Economics and Forest Resources, University of Washington, Seattle, Washington

When asked how much it cost to own and operate his yacht, the *Corsair*, the financier J. Pierpont Morgan is reported to have responded, "If you even have to ask the cost, you can't afford one." Applying Mr. Morgan's dictum to the management of our national forests, we might reasonably pose the question: If society knew what its national forests were costing, would it even want to keep them?

The national forests provide a good many benefits, a fact repeatedly emphasized by their Forest Service managers and by vested interests which find existing arrangements to their advantage. We should not, however, let ourselves be distracted by self-serving arguments. To do so is to ignore the fundamental question of whether if *all the costs and benefits* of our national forests and their present management were properly counted, they would be a *net benefit* to the American people.

It is not the purpose of the discussion to answer that question. Rather, I would merely like to identify some of the costs that are commonly neglected in evaluating the net benefits the national forests provide to the American people. The specific costs I would like to discuss are the carrying charges on forest land and timber inventories which are neglected in planning national forest management, or in evaluating the net benefits the forests produce. This by no means exhausts all the hidden costs associated with the national forests. It does cover some of the more obviously burdensome ones.

A disclaimer

First, I would like to interject a disclaimer. It is one thing to hold personal opinions about issues; it is quite another to analyze the implications of various management alternatives objectively so people can make informed choices. Part of the training of an economist is to distinguish between the *analysis* of various alternatives and the *advocacy* of a specific alternative that may be best to one's liking. I am a strong advocate of analyzing issues, and getting the truth out so that people can make informed choices. I also have prejudices which are no secret. I think the present management of the national forests is irrational, inefficient, and a burden to society. Even more distasteful, it redistributes income from the relatively poor to the relatively rich. The average U.S. taxpayer could do quite a bit with the timber sales receipts he could get if he sold some of his timber. Instead, we seem to have these lands locked up as playgrounds for the relatively wealthy. This is not a majority opinion.

I also recognize that pride in our national forests and their Forest Service managers, like political party loyalties and nationalism, are collective rather than individual preferences. The general public might very well choose to continue having the national forests managed as they currently are even if it knew it was sacrificing the advantages of alternative forms of management. Preferences of these kinds are not unlike those described so expressively by economist Joseph Schumpeter: "Some people prefer socialist bread to capitalist bread even if it has mice in it."

Many countries have opted for public ownership and management of natural resources because they felt they were God-given and should, therefore, be treated differently from other kinds of capital assets. If a country wants to sacrifice income for beliefs of these kinds, the economist should not complain. His primary responsibility is to estimate fairly what the sacrifices are.

The concept of carrying charges

Imagine a cattle rancher who, in determining the profitability of his beef-producing operation, disregarded the value of his land and his cattle herd. Quite possibly, his operation would appear to be profitable. In fact, if he counted all his costs, he would be losing money. He always has the opportunity to sell his land and herd, put his money into a savings account or bonds, and collect the interest from his deposits or securities. Interest earnings from these options are legitimate costs to charge against his ranching.

Suppose a small operator did not count the cost of his plant and equipment, or carrying charges on his lumber inventories, in determining the profitability of his operation. It is not likely he would survive long in the competitive lumber market—unless, of course, someone was willing to provide him with generous subsidies.

The situation is no different for a tree farmer. His land and timber inventory have value. He always has the alternative of selling, and investing his money elsewhere. Unless the net receipts from his operation, including interest charges on the value of his land and timber inventory, are positive he is losing money.

The supply of beef, the supply of lumber, and the supply of timber produced on private lands all depend on the costs associated with land and capital investments. If these costs are not counted and a producer sells his output in a reasonably competitive market and survives, then he must be receiving a subsidy, either open or hidden.

The costs of neither land nor timber inventories are counted in the management of the national forests, and this does represent an enormous subsidy to the Forest Service and some of its customers. If these costs were more widely known, it is not obvious the Forest Service would be able to continue managing the national forests on the basis of the "sustained-yield nondeclining evenflow" concept. (In a society accustomed to acronyms, perhaps we could call their model SYNDEF. It seems appropriate. It is a *sin* against rational economics, and many of its proponents are *deaf* to any alternatives.) Surely, it is in the interest of efficient resource use and sound public finance that these costs be made known. This would help clarify the alternatives and enable the public to make more informed choices.

Estimates of carrying charges on land and timber

What are the magnitudes of the carrying charges on land and timber inventories on the national forests if they were counted? Without careful study, estimates of these kinds are at best impressionistic, but impressions are sometimes better than nothing. The Forest Service reported in its *Timber Outlook*, published in 1973, that there were 17,427,000 acres of commercial forest lands on the national forests in Oregon and Washington as of January 1, 1970. The softwood sawtimber inventory on these national forests was 388 billion board feet (International 1/4-inch log rule). This was 46 percent of the softwood sawtimber inventory in the two states. Needless to say, the Forest Service is in a position to significantly affect the forest products economy of the two states.

It is not unreasonable to assume the 1970 estimates of commercial forest land acreage and timber inventory approximate what exists today. If land were valued at $25 per acre, and softwood sawtimber at $50 per thousand board feet, both reasonably conservative estimates, the total value of national forest timber and forest land would be nearly $18 billion. If the national forests in the two states were ranked among the "Fortune 500" list for 1974 in terms of assets, they would be in third place, between General Motors and Texaco.

The $58 million in "in lieu" payments distributed to Oregon and Washington counties in 1974 would indicate receipts from timber sales of $232 million. Timber management expenditures were slightly more than $22 million. Approximately another $18 million was spent on timber access road construction. Combining these estimates provides a crude estimate of net cash flow of $192 million, approximately 1.1 percent of the value of land and timber inventories. This is not a measure of rate of return, because no costs have been deducted for land and timber inventories.

At a mere 6 percent rate of interest, returns on assets worth $18 billion would be $1.08 billion per year. If these costs were deducted, the national forests in Oregon and Washington would be losing almost a billion dollars a year. The cost of sustaining the SYNDEF approach to timber harvest scheduling is obviously rather high. It is small wonder the Forest Service is not anxious to make these costs known. This raises the question, of course, of whether it should. There are precedents for suggesting it should. Taxpayers do have to pay interest charges on their national debt. Why shouldn't they be receiving the equivalent of interest earnings on their national forest assets?

A similar, and more detailed estimate of the net returns from the management of all the national forests was reported recently in an article by Dr. Marion Clawson, an economist; former director of the Bureau of Land Management; and until his recent retirement, president of Resources for the Future. Dr. Clawson's article, titled "The National Forests," was published in *Science* (2-20-76), the journal of the American Assn. for the Advancement of Science. He concluded the national forests were a valuable asset, but they were being poorly managed and were unproductive. Dr. Clawson estimated the value of the national forests to be \$42 billion. This is \$11 billion more than Exxon which, with \$31 billion in assets, topped the Fortune 500 list for 1974. In contrast to Exxon, which had a net income of \$3.1 billion in 1974, the national forests had a cash deficit of \$2 million, according to Dr. Clawson. (And 1974, by the way, was a pretty good year. The deficit is generally much larger.)

After imputing values for recreational and wilderness use and for water, and deducting carrying charges on land and timber inventories calculated at 5 percent, Dr. Clawson concluded the national forests ran a deficit of nearly \$2 billion. This is about \$9 for every man, woman, and child in the country.

Dr. Clawson's conclusions will undoubtedly attract much attention, and will be criticized by those who think his estimates are either too high or too low. As interesting as these debates among the professionals may be, we should not permit them to distract us from the substance of his conclusion: The national forests are currently a dead-weight loss to society. In my opinion, this is a correct conclusion. The academic and professional quibbling about his estimates will merely help refine the accuracy of the losses.

Professor Jack Hirschleifer, an economist at UCLA, confirmed these conclusions at a symposium, "The Economics of Sustained Yield Forestry," held at the University of Washington in November 1974. Professor Hirshleifer concluded that, given the current management of national forests, the American people might be better off if they simply gave them away.

Some communications problems

These conclusions stand in sharp contrast to the reputation the Forest Service has established with the general public as a socially responsible organization which practices "good conservation" in the management of the national forests. A commonly held view is that if it weren't for the Forest Service, there wouldn't be any forests. They would be ravaged by profit-seeking businessmen.

The problem, of course, lies in the fact defenders of SYNDEF have succeeded in vesting the concept with such laudatory semantic connotations it is difficult to evaluate critically to any except trained economists. Note the intuitive appeal of the logic. People want perpetual wood supplies and stable communities and industries. SYNDEF, they are told, is a means of achieving these economically and socially desirable objectives. To be against SYNDEF is to be against desirable objectives and is irresponsible.

Trying to explain to the layman, or to a professional who is committed to SYNDEF, that it is bad economic reasoning is not an easy task. Economics is complicated. If an economist omits details from his argument in an attempt to clarify the substantive issues, he is criticized for being too abstract. If he introduces detail, he is criticized for being too complicated. If he cannot do one or the other, then he simply cannot win. The situation has been aptly described by Luther Hodges, Secretary of Commerce during the Kennedy Administration: "If ignorance paid dividends, most Americans could make a fortune out of what they don't know about economics. Hardly one person in twenty has the sketchiest idea of how our economy functions. America may be the greatest industrial power on earth, but for all that most of us know, our greatness might depend on growing breadfruit trees." Or, we might add, sitting on all that national forest timber.

It is an unfortunate fact bureaucratic and other interests trade on this kind of ignorance to perpetuate and promote their own advantage. Try to get a Forest Service official to defend SYNDEF on the basis of logic and factual evidence. Their principal defense is a dogmatic belief that economic criteria (correct economic criteria, that is, not the ACE and RAM models used by Forest Service economists) and the price system do not work in forest management. To use economic criteria, they seem to believe, would expose the country to the risk we would run out of trees. No responsible agency could do that. The belief that a market economy, like Mother Hubbard, can run out of supplies is to fail completely to understand the workings of the price system.

It is interesting to note that timber producers in the South have not discovered the price system does not work in forestry. They harvested an estimated 330 billion board feet of sawtimber between 1950 and 1970. The inventory in 1950 was estimated to be 357 billion board feet. In 1970 it was reported to be 438 billion board feet, an increase of 81 billion board feet. Most of the forest land in the South is privately owned (95 or 96 percent) and managed, presumably, under market criteria and the price system.

What are the alternatives?

Even superficial examination of SYNDEF indicates it is not compatible with a market economy. It is not even compatible with a socialist economy. SYNDEF is a product of feudalism where communities were self-sufficient, wood was the primary source of fuel and building materials, and the peasants had rights of common use of the forest to cut their wood needs. There was a possibility in that environment of overcutting if regulations were not imposed. To use the phrase made popular by the biologist Garrett Hardin, there was a danger of a "tragedy of the commons." Contrary to claims that SYNDEF is a means of promoting community stability, it should be recognized as a result of a condition of economic stagnancy.

When communal rights were abolished and private (exclusive and transferable) rights were created, the threat of timber depletion was eliminated and the foundation was laid for the use of the price system in forest management. Unfortunately, it was not adopted by most of the forestry profession. It still has not been adopted in public forestry even though its logic has been demonstrated and it has proved its worth in the South.

It took centuries for Western societies to overcome the stigma associated with the use of an interest rate in guiding economic decisions. Even the socialist economies are now finding it useful, although they usually call it something else. The Forest Service has been in existence only about three-quarters of a century. Perhaps we should be patient.

DISCUSSION PERIOD

QUESTION: I was interested in your assumptions about the value of national forests. A member of the Administration recently suggested we place a value on all our national assets, apply those values against the national debt, and cancel it out. I wonder if this type of paperwork shuffle would operate in the real world. Who would buy the timber at the value you have assumed?

DOWDLE: I imagine many would be willing to buy it as a basis for establishing new industry. We have enormous forest resources which could provide raw materials for growth in the forest products industry. We really are not making full use of it. The rate at which it is being used is pitifully low. The statistics for Washington and Oregon show the rate of timber inventory turnover in the public sector is 1.4 percent. Given the overmature condition of that timber, that is much too low. If this rate were to be doubled—to 2.8 percent, which would still be low by rational economic criteria—output in the two states would be increased by about 7.5 billion board feet. The American people are clearly being deprived of the revenues and product benefits they could get from selling that timber. There are also world markets.

One thing continually overlooked in most arguments about national forest timber is that it is essentially an argument among users. What about the American people, who have owned that timber for years? They should welcome the opportunity to sell some of it and get a return from their investment. I do not think it is incorrect to place a value on an asset if that asset is being used to produce a flow of services. Efficient management requires this be done.

QUESTION: What is the turnover rate in the private sector?

DOWDLE: The turnover rate in the private sector is about 4.5 percent. It depends somewhat on how the private sector is defined.

QUESTION: What do you suggest we do to change the policy of the federal government?

DOWDLE: I am not really quite sure. It is a political, as opposed to an academic, issue. I must confess, I haven't been able to change much. I suppose people should keep after them, keep asking them questions. Most public foresters are, however, extremely evasive. They are experts at muddying waters and begging questions. I believe it may take a crisis to initiate reform. Given the current timber supply situation in the Pacific Northwest, with much high-priced timber under contract, there will be pressure for change. Something may give.

QUESTION: What is the mission of the Forest Service regarding economic considerations? It is required to take economics into consideration?

DOWDLE: The legislative mandate of the Forest Service does not make clear what it is supposed to do. The Sustained-Yield Multiple-Use Act of 1960 is a monument to ambiguity. It is a highly permissive piece of legislation. If it is not clear what an agency is supposed to do, then it can do many things—some good and some bad—and remain consistent with its legislative mandate. The Forest Service is not breaking the law. I do think we have 75 years of forestry legislation based on the notion of managing forests in a feudal society. The 1897 Organic Act, which is creating some difficulties now, is a good case in point. It is the kind of law a feudal lord would write if he wanted to keep profit-seekers out of his forest.

One of the laws under which the Forest Service operates, the McSweeney-McNary Act, says the agency is supposed to develop economic considerations relevant to forest management. The agency has produced numerous studies. If an economic analysis is performed and the conclusions are not to its liking, the Forest Service seems to reach the conclusion economic analysis is not relevant. A good example was its forest regulation study. The procedure is somewhat like performing surgery with a rusty tin can lid. If the operation isn't successful, the conclusion is made that surgery is no good.

Many of the so-called economic analyses that are done by Forest Service economists are very disappointing. Their use of the so-called "allowable cut effect" is a good case in point. It badly confuses the whole issue of investment analysis. But the Forest Service is able to perpetuate the belief it is performing economic analyses and the public is receiving excellent rates of return on certain kinds of investments. This is incorrect.

Timber RAM is another case in point. Some of the constraints in the model are economically irrational. Unfortunately, there is a tendency for people to suspend their common sense when confronted with simulation, linear programming models, and computer output. These "black box" models are being sold to the general public as rational planning tools. If you open them up and look at their details, you find much the same old nonsense inside—the sustained-yield, nondeclining, evenflow concept of timber harvest scheduling.

QUESTION: In the last large sale for which I have any information, the timber went for $0.50 per thousand and there was no value placed on the land. That was a private sale—a private buyer and a private seller—and the timber was on easily logged land. Forest Service timber in many cases is on very undesirable land, very hard to log. In fact, some of it might not be salable at all. So I question your figures, your assumptions, as to the value of the land and of the timber. Most of the large timber stands were obtained essentially without a dollar value. For example, the railroad grants contain much of the timber in the West. I again question the logic of an official value on the land and the timber.

DOWDLE: Timber is an economic commodity. Timber inventories represent capital—you can run it down through harvesting; you can build it up through investments in intensive forest practices. But you have an inventory of timber which you are using to produce a flow of logs.

Most appreciate that a sawmill operator has some notion of an optimal lumber inventory to cover his sales; he has a notion of an optimal log inventory to plan his production. These inventories represent capital assets, and their values are necessary information to manage them efficiently.

All I am suggesting is that the Forest Service should be using economic criteria—the price system and, specifically, the interest rate—in its planning. If it uses the interest rate correctly, it needs to know something about asset values. Timber production cannot be planned in an economically efficient manner unless some rules, some criteria, are used. The Forest Service is presently managing capital without using the rules of optimal capital management. This is the heart of the problem.

Now, you may quibble about the price. I stated specifically that I priced out the National Forest lands in Washington and Oregon at an average of $25 per acre. I have a feeling you could sell quite a lot of forest land at that price. I priced out the timber at $50 per thousand board feet. I do not think those numbers are unreasonable. Perhaps I should take the position of some of my colleagues in the economics department. When somebody questions their numbers, asks why, they respond, "Why not?" If you don't like the numbers, use your own, but the point is you should use some.

Timber does vary a great deal in value. There is no question about it. But if you do not value it and apply rational economic criteria to these value differentials, consider the position in which you place yourself. Can you conceive of the coal industry going down to the deepest, narrowest veins of coal and digging? No. The best veins are mined first. That is the rational way to proceed. In forestry, on the other hand, one can drive through miles of beautiful stands of timber to get back to some ridge where the timber is small and rotten. The intent is to log it and get the forest land back in production.

I am reminded of my grandfather and his apples. When I wanted to eat one, he would always tell me, "Son, take the rotten ones first." We'd cut the rot out before we ate them, of course. At the end it seemed as if we had eaten rotten apples all year. Similarly, the Forest Service wants us to drive through stand after stand of beautiful timber to get at the rotten material. This is nonsense. It is not against the law. My point is that the laws are perhaps in need of changing. The industry will continue to have problems with our forestry laws as long as they are on the books. They open the door for raids on the public domain. Many users benefit without paying for what they get.

As a taxpayer, I would like to note that the forest products industry pays for every stick of timber it takes. I get a return from timber that is sold to the industry. What return do I get as a taxpayer (stockholder) from timber capital used as wilderness? Nothing. These issues should be taken into account. More specific legislation based on rational economic criteria could make the public more aware of these issues.

Exhibitors
1976 Sawmill & Plywood Clinic
Portland, Oregon

The many practical and innovative ideas offered by the 177 companies that exhibited at the 1976 Sawmill & Plywood Clinic added immeasurably to the Clinic's success. Registrants had the chance to speak with many exhibitors' representatives, receiving information focused on individual problems or simply gathering ideas for modifications under consideration. As registrants made their way from booth to booth, the opportunity to share ideas with other mill operators was extended beyond the Clinic auditorium.

Unfortunately, the specific information provided by the exhibits cannot be recorded here—but testimony from the many mill people viewing the show confirms its importance. The Clinic sponsors appreciate the support shown by the companies exhibiting in Portland. The following companies were represented at the 1976 Portland Clinic:

Advanced Combustion Systems, Inc.
Albany International Industries, Inc.
Allen-Bradley Company
Aloha Metal Fabricators, Inc.
American Manufacturing Company, Inc.
American Sheet Metal
Applied Theory Associates, Inc.
Archer Blower & Pipe Company
Argo International
Armstrong Mfg. Co.
Artek Industries, Inc.
Auto-Chip Corporation
Automatic Valve Corp.

Bell Industries
Bigelow Machinery Company
Bird-Johnson Company
Black Clawson, Inc.
Bohler Bros. of America, Inc., U.S. Knife Div.
Boyd Industrial Rubber
Brunette Machine Works, Ltd.
Buchanan Fluid Systems
Bumstead & Wolford Co.
Burelbach Corporation

CAE Machinery Ltd.
Carothers Sheet Metal
Cascade Plating Company, Inc.
Chip-N-Saw, Inc.
Clarke's Sheet Metal
Climax Mfg. Co.
Coastal Machinery Co., Inc.
Coe Mfg. Company
Conveyco Manufacturing Corporation
Cothrell Machinery Sales, Ltd.
Cutler Hammer

Davis Industrial Products Company
Decision Dynamics, Inc.
Disston, Inc.
Dunhill of Portland, Inc.
Dupar Dynamics Corporation
Durand Machine Co., Ltd.

Electrospace Systems Inc.
The Ellison Company
Energy Products of Idaho
Esco Corp.
Extremultus, Inc.

Farwest Equipment & Controls Corporation
Fluid Air Components, Inc.
FMC Chain Division, FMC Corp.
Furane Plastics
Futerized Systems, Inc.

Gaskell Company, Inc.
Gates Rubber Company
Gear Reducer Sales Company
General Bearing Company
General Electric
General Kinematics Corp.
Gerlinger Power Products Company
Gerrard Sales & Service
Glidden Durkee, Div. of SCM Corp.
Globe Machine Mfg. Co., Inc.
Gruendler Crusher & Pulverizer Co.

Habasit Belting, Inc.
Hanchett Company
Harvey Engineering & Mfg. Corp.
Walter E. Heller Western Incorporated
Hewlett-Packard Company
Hitachi America
R. Hoe & Company, Inc.
Howard-Cooper Corporation

Industrial Electric Service Company
Industrial Machine Co.
Industrial Machinery Erectors Co.
Interlake, Inc.

International Enterprises
of America
Irvington-Moore

Jacksonville Blow Pipe, Div.
Montgomery Industries
International
Jeddeloh Bros. Sweed Mills
Jeffrey Power Transmission

Keystone Machine Works
L. F. R. Knudson Company
Kockum Industries

Lamb-Grays Harbor
Company, Inc.
Lebanon Machine Works
Lewis Electronics, Inc.
Lincoln Engineering
Lloyd Controls
L–M Equipment Company, Inc.
Los Angeles Brush Mfg. Corp.
Lumber Systems, Inc.

Mac-Fab Mfg., Inc.
Machinery Sales Co., Inc.
Manitowoc Shipbuilding, Inc.
Marathon Le Tourneau
Company
Mark 50 Machinery
Martin Sales Engineering Corp.
H. C. Mason & Associates
Maxco
McCarthy Products
McCracken Machinery Sales, Inc.
McDonald Mfg. Company
McDonough Mfg. Company
McGuire Bearing Company
Metal Detectors, Inc.
Micro Switch,
Divison of Honeywell
Mill-Tech
Milwaukee Crane & Equipment

Mo Do Mekan, Inc.
Moore Oregon–Canada
Morvue, Inc.

National Builders Hardware Co.
Newman Machine Company, Inc.
Nicholson Manufacturing Co.
Nopak Northwest
North American Controls
Northwest Chain-Flowsystems

Opcon, Inc.

Pace Products Inc.
Pacific Processing Equipment
Company
Pacific Rubber of Oregon, Inc.
Pacific Saw & Knife Co.
PED Manufacturing Ltd.
Peerless Div.–Royal Industries
Platt Electric Company, Inc.
Plywood Equipment Sales
Pneumatic Systems, Inc.
Portland Chain Mfg. Company
Portland Iron Works
Portland Machinery Co.
Power Transmission Products
Premier Gear & Machine Works
E. V. Prentice
Process Computer Systems
Proctor Sales, Inc.
Protective Coatings

Rader Pneumatics
Raute Inc.
Ray Burner Company
Record-a-Call Telephone
Answering Systems
Reid-Strutt Company, Inc.
Reliance Electric
Rens Manufacturing Company
Rexnord Inc.
Ringfeder Corporation

Rogers Machinery Co., Inc.
Rolie's Machine Works, Inc.
Ronan Drying Systems, Inc.

Salem Equipment, Inc.
Schurman Machine Works, Inc.
Seattle Construction Co.
Signode Corporation
Simco-Ramic Corp.
Simpson Timber Company
Smets Machinery
Sprout, Waldron
& Company, Inc.
Square D Company
Star Rental & Sales
Stetson-Ross Machine Co., Inc.
Strom Industries
Superior Plywood
Machinery, Inc.

Tallman Machinery
Company Inc.
3M Company
Timber Conversion, Inc.
Top Line Equipment Company
Totem Equipment Co.
Trienco, Inc.
Tyrone Hydraulics, Inc.

Unico, Inc.
United Tool Company
Universal Plastics Company

Wagner Electronic Products
Ward Systems, Inc.
Wellons, Inc.
Westcon, Inc.
Western Fluid Power
Western Machinery Corp.
Westwood Structures

A. J. Zinda

Index

Abrader arm maintenance, 114
Agricultural lands, preservation of, 280, 282
Air cannon. *See* Big Blaster air cannon
Air compressors
  coolant-lubricant functions, 231
  dirt in, 232-33, 234
  new coolant, 233-35
  types, 230-31
Air pollution, from chipper, 165
Automation, 58-72
  board feet per cubic foot with, 58, 72
  computer readouts, 66-69
  design requirements, 60-61
  locating malfunctions, 69-70
  logic systems, 141-42
  operator importance, 266
  personnel and suppliers, 61, 70, 71, 86
  See also Computer; Computerized edger

Baker decision, 279, 281
Bandmills, 76, 78, 123
  axial vibration, 181
  belt drive types for, 183-84
Bandsaws, 118, 121
Bark, conveying system for, 94, 95
Barker, 76
Barking-bucking line, 83
Barking system, for small logs, 80
Belts
  basic equation for, 174
  centrifugal force effects on, 178
  criteria for, 173-74
  double pull compared to Sweeds, 134-35
  drive types compared, 182-84
  elasticity and tensile strength, 176-77, 178
  equation for effective pull, 176
  flat synthetic, 173-85
  shock absorption by, 176-77
  speed of, 179
  troughing, 134
  unit weight of, 177-79
  See also V-belts
Belt transmission drives, 173-85
  bearing load, 177-80
  comparison of types, 182-84
  constant coefficient of friction, 174-75
  flat synthetic belts, 182-84
  static shaft load, 183-84
  synthetics introduced for, 175-76
  vibration effects on, 180-81, 183
Big Blaster air cannon, 236-41
British Columbia, 164, 242, 243, 245
Bucking
  computerized, 49-51
  delays and problems with, 76
  small-diameter long logs, 114-15
  speeds, for small logs, 79, 80
  station, 62, 63
  system, for small logs, 80
Bulk storage containers, 236-41
Bundle log lifts, 112-13
Bunk design, 119-20
By-product conveying system, 93, 94-95

Cant
  flippers, 196, 197
  knee turners, 120-21
  separators, 92
Cant-producing facility, 88, 91
Carriage bunk design, 119-20
Cedar, 123
  hog fuel conveyance, 134
  shavings, value of, 244
Chairs
  chain tighteners for, 113
  misalignment of, 110-12
  spacing of, in log hauls, 102-3
Chipper
  belt drive types for, 182-83
  cast steel discs for, 167, 168
  design of for chip quality, 166-71
  disc speed, 166
  gravity-feed, disadvantages of, 163, 164
  horizontal-feed, 163, 170
  knife-to-anvil clearance, 169, 170
  machine size, 165
  pitch buildup on, 138
  protection from metal damage, 171-72
  selection of, 162-66
  set-up and maintenance of, 135-38
  top or bottom discharge, 165-66
  See also Wastewood chipper
Chipper-canter, 91
Chipperheads, 64, 116-17
Chipping
  fine reduction, 164
  pine or fir, 138
  trim ends, 164
Chips
  avoiding fines and oversize, 135-38
  bulk storage containers for, 239
  conveying system for, 94, 95
  longer lengths of, 164, 165
  quality of, 166-71
  replaced by boards, 160-61
  See also Sawdust chips
Clawson, Marion, 292
Coastal zone management, 278, 284
Columbia Region Assn. of Governments (CRAG), 280, 287

Communication
importance of, 259-60
improvements in, 260-62, 263
for quality control, 202-5 *passim*, 218
Computer
central processor unit, 141-42, 149, 151
control of veneer stacker, 145-48
control vs. hardwired control, 144-45
design factors, 142-43
instructions, 142, 143, 153
for inventory control, 71-72
for land-use planning, 275
programs, 142, 143, 151
readouts, 66-69
as a universal control element, 148
voltage requirements, 150
Computerized bucking
advantages of, 56-57
production increases from, 49-51
Computerized edger
board measurement, 156-57
boards replace chips, 160-61
decisions by operator, 155-56
profits from, 160-61
sideboard recovery increased, 154-61
volume and price relationships, 157-59
Conveyors
improvements in, 132-35
vibrating, with chipper, 163, 164
Coolant-lubricants, 231, 233-35
Costs
increases in, 35
of kiln operators, 250-51
of national forest management, 288-298
quality control, 190-92
of timber, 297
See also Fuel costs; Lost-time dollars
Cranes, 39, 41, 44, 45

Debarker, 88, 91
Debarking inefficiencies, 113-15
Deck problems, reducing, 105-13, 208-13
crossed and tangled logs, 107-8
side lifts, 110-12
Douglas fir, 76, 84, 196, 244
Drive efficiency, 173-85
Dry kiln design, and woodwaste recovery, 242-57
Dry-land log handling, 35-47
scaling methods, 43-44
storage area, 46
unloading machinery for, 39-43

Edgers, 79, 85
accuracy improved, 195-97
belt drive types for, 182
decisions made by operator, 155-56
problem areas, 125-27
See also Computerized edger
Environmental protection, 276, 277-78

Fiber production, 273-74, 276
Fir, 138. *See also* Douglas fir
Flakeboard plant, quality control, 216-228
Flat synthetic belts, 173-85
Forest product industry, 262-64
Forest Service. *See* U.S. Forest Service
Forests, optimum growth from, 73-74
Forklifts, 35-36, 39, 45, 46
Fuel costs, 242-47
for gas/butane heated kilns, 245-47
in million Btus, 242-43
of shavings, 243, 244-45

Gamache, Adrian, 274, 278
Graphs, for quality control, 197-99, 200, 203
Grapples, 39, 41, 88
Growth rings, and chip size, 123, 124

Handbooks, 205, 206, 208
Headrig
problem areas of, 116-21
quality control for, 193-95, 200
Hemlock, 245
Human engineering, 265-72

Industrial design, 269, 270-72
Infeed
chain, at edger, 160
deck, 49, 54, 55
to resaw, for small logs, 91-93
Inventories
computerized, 71-72
costs of in national forests, 288-89, 290-92

Jackson, Cliff, 76, 79

Kilns, dry and woodwaste recovery, 242-57
Kuhn, Stan, 226

Land Conservation & Development Commission (LCDC), 279, 280-81, 283-84, 285, 286-87
Land-use legislation, 278, 281-82
Land-use plans, state, 278-86
coastal zones, 278, 284
comprehensive plans, 279, 281-83, 286
economic considerations, 282-83, 287
exclusive farm-use zones, 280
flood plains, 280, 283-84
urban growth boundary, 281
zoning laws, 281-82, 284-85
Land-use plans, U.S. Forest Service, 273-78
impact on private industry, 274-75
"marginal" designation, 276
"special interest" designation, 275-76, 283
Large-log mill, modified for small logs, 73-85
Log
cutoff saws, 114-15

diameter, 66, 72, 74, 78, 79, 84-85, 88
handling, dry-land, 35-47
haul efficiency, 102-5
infeed, 102-5
loaders, 118-19
sprinkling, 209-13
storage, 46, 209, 211-13
transfer deck problems, 105-13
unloading, 39-43
Logic function control, 141-53
Logs, small. *See* Small logs
Lost-time dollars, 100-40
chippers, 135-38
conveyors, 132-35
drop sorter design, 138-40
saw speeds, 121-25
unscramblers, 130-31
Lubricants, 231, 233-35
Lysholm, Alf, 229

Machinery, monitoring of, 193-95, 197-99
McSweeney-McNary Act, 296
Management
booklets for employees, 260-62
computer data useful for, 66-68
quality control by, 187-88, 201-3, 217-19
women and minorities in, 263-64
Man-machine system, 266-69
Manuals, for machine operators, 205-08 *passim*, 226-27
Market prices, 157-59
Metal detection, 55, 171, 220, 222, 224
Mill
large-log, 73-85
yard, profitability of, 36-37, 47
Moisture content, 221, 222, 245
Monongahela National Forest, 262, 263

National forests
land-use plans, 273-78
timber harvest reduction, 262
value of land and timber in, 291, 295, 297
National forests, costs of, 288-298
acres of commercial forest lands, 291
"allowable cut effect," 296
carrying charges on inventories, 288-89, 290-92, 297
legal aspects, 296
net returns, 291-92
RAM model, 293, 297
"sustained-yield nondeclining evenflow" concept (SYNDEF), 290-94 *passim*

Oil coolants, for air compressors, 229-35
Operators, 250-51, 266-69
Oregon
acres of commercial forests in, 291
land-use planning in, 278-86
price of national forest lands in, 291, 298
Oregon Coastal Conservation & Development Commission, 284
Organic Act (1897), 296

Personnel
in automated systems, 61, 70, 86
back-up operators, 95-96, 98
communication among, 259-62
human engineering approach, 265-72
optimum use of, 258-64
political awareness of, 262-63
Pine, 76, 84, 138
Pitch, 138
Planer, 93-94, 213, 214
Pollution, 165, 209, 212-13, 281
Power transmission, belting for, 173-85
Prices, of boards related to volume, 157-59
Profits, 160-61. *See also* Lost-time dollars
Public lands. *See* Land-use plans

Quad line, 79, 80, 83
Quality control program, 186-228
laboratory tests, 224-225
log deck to dry kiln, 208-14
personnel, 192-93, 199, 201-5 *passim*

Raw material
diversity of, 87-88, 91
estimating future resources, 78-79
Raygo-Wagner Lumberjack, 36, 39
Redwood, cant flippers for, 196, 197
Resaw linebar, 131-32
Resaw system, for small logs, 88, 91-92
Resin mixing, 222
Roll cases, 160
chromed, 135
fluted, 108, 110, 115

Sawdust chips, sliding out of gullet, 121-25
Sawmill Clinic, 72, 121
Saw settings, for trimmer, 128-30
Saw speeds, 121-25
Saw teeth, for sawdust, 123-25
Sawyer's cab, 116-118
Scaling methods, 42-43
Scanning system, computerized, 58, 60, 62, 63, 64
edger optimizer with, 154-55, 156-57
visual judgment compared to, 96-97
Scragg mill, 64
Shadow lines, 118
Shavings
Btus per pound, 244, 254
conveying system for, 94, 95
cost of for fuel, 243
value of compared to other fuels, 244-45
Sideboard
measurement, 156-57
recovery, with computerized edger, 154-61

Side lifts, design of, 110-12
Silicones, for coolant-lubricants, 233-34
Small-log processing, 86-99
  back-up operators, 95-96, 98
  by-product system, 94-95
  chipper-canter, 91
  infeed to resaw, 91-93
  large-log mill, modified for, 73-85
  raw material variations, 87-88, 91
Small logs, 73-74, 78, 118
Sorting systems, 48-57
  construction of bays, 52
  diversity of log load, 87-88
  drop sorter design, 138-40
  elevated feed deck with, 49
  metal detection in, 55
  speed and accuracy of, 48-57
  tangled logs, 47
  upgrading included with, 44-47
Sprinkling logs, to improve recovery, 209-13
Spruce, 245, 257
Storage containers. *See* Bulk storage containers
Stud mills, 63-64, 205
Sustained-Yield Multiple-Use Act (1960), 296
"Sustained-yield nondeclining evenflow" (SYNDEF), 290, 291, 292, 294
Synthetic flat belts, 173-85

Timber
  costs of, 297
  supply problem, 58, 59
  surveys of, 75, 291
Timber inventories, costs of in national forests, 288-89, 290-92, 297-98
Timber Outlook, 291
Time lost. *See* Lost-time dollars
Totem Sateko sorting system, 44, 48, 51-55
Trim ends, chipper for, 164
Trimmer, 93-94
  with memory system, 93
  saw settings, 128-30
Troubleshooting, 153
Troughing belts, 134

Universal logic element, 141-43
Unloader arms, design of, 110
Unloading logs, 39-43
Unscramblers
  computerized edging system, 155-56
  design improvements, 130-31
Upgrading, 44-47, 48-57
U.S. Environmental Protection Agency, 213
U.S. Forest Products Laboratory, 249
U.S. Forest Service, 58
  economic considerations, 293-94, 296
  harvest per year in Region VI, 274
  land-use plans, 273-78
  legislative mandates, 296
  See also National forests, costs of
U.S. Plywood Corp., 73

Vane air compressors, 230, 233
V-belts, 174, 175, 176, 185
  bearing load of, 177, 179, 180
  compared to other drive types, 182-84
  horsepower delivered by, 179
  vibration effects on, 180-81
Veneer business, 70
Veneer stacker, computer vs. hardwired control of, 145-48
Vibrating conveyor, 163, 164
Vibration, effects of on belts, 180-81, 183, 184, 185
Vibrators
  electrical and pneumatic, 240
  industrial, 236, 240-41
Voltages, for computers, 150
Voting, and forest product industry, 262, 263

Washington State
  acres of commercial forest in, 291
  land-use plans, 278
Wastewood chipper, selection of, 162-72
Wastewood chip types, 162-65
Wastewood-fired boilers, 252, 253
Waterfalls, design of, 131-32
Water log storage, 209, 211-13
Water pollution control standards, 209, 212-13
Western Environmental Trade Assn. (WETA), 277, 284, 285
White fir, 76, 84
Willamette National Forest, land-use planning for, 275-78
Willamette River Greenway, 283-84
Willamette Timbermen, 277
Women, in executive positions, 263-64
Woodwaste recovery, and dry kiln design, 242-57
  requirements of woodwaste for, 254-55
  value of woodwaste, 244-45